All rights reserved. Printed in the United States of America. No part of this publication may be reproduced, stored in a retrieval system or transmitted in any form or by any means, electronic, mechanical, photo copying, recording, or otherwise, without the written permission of the publisher.

Note: This is a work of non-fiction. Some names/characters were changed and placed are either the product of the author's imagination, or are used fictitiously, in order to not have any resemblance to actual persons, living or dead, business establishments, events, and certain locales is entirely coincidental.

Copyright © 2017 by Reva McManis

ISBN – 13:978-1981272259
ISBN – 10:1981272259

Printed in the United States of America

For all general information contact Createspace

Cover Done By: Reva McManis

I would like to acknowledge all individuals for their permission to print their stories that were penned anonymously (or) that are public domain or were written by them:

    1.) Printed By Permission Of Barbara Ellen McKittrick (Author) @ 2017

Contents

Introduction: Welcome To "Bigfoot Madness"

Chapter 1: Started "Bigfoot Madness" In 2016   5

Chapter 2: The Colony   9

Chapter 3: Unexplainable Things   16

Chapter 4: Evidence   20

Chapter 5: Photography   27

Chapter 6: My Baby Pictures   31

Chapter 7: Bears   35

Chapter 8: The Rock Experiment   36

Chapter 9: The Unknowns Of Research   39

Chapter 10: Powder Appears   43

Chapter 11: September 18, 2015 Expedition   45

Chapter 12: They Just Vanish   52

Chapter 13: Babies   58

Chapter 14: Special Photos   66

Chapter 15: More Evidence   88

Chapter 16: The "Dogmen" Encounter   100

Introduction

Welcome To "Bigfoot Madness"

Since my first sighting of Bigfoot in March, 2013 – I have had numerous expeditions which have resulted in some of the strangest occurrences; if you did not know me personally, you most likely would think I had lost my mind and entered into my own world of "madness". I assure you the photographs (are real), the stories (are real), and the strange noises/sounds (are real), and the details to each occurrence (are real) from the local area of Carroll County, Kentucky in which I live.

The "Bigfoot" World of Researchers are now coming to the point of many new findings. Rocks that show etchings of unknown origin, visual sightings in the sky of unknown craft, strange pictures of E.T.'s with elongated skulls, colonies close to one another of Bigfoot, and the strange stick formations left by them.

This year, I felt I have strengthened my writing abilities, my research abilities, and also have increased my level of courage for some of this research. I have thoroughly enjoyed putting on the First Annual Bigfoot Conference in our local area and have certainly tried to lecture to schools on this subject matter. Our history and science will be written again but with a slant to the "stars" and our visitors. I, like everyone else, was taught in school – the standard biological science with regards to our Earth, the Constellations, Planets, and our abilities to travel in space. Little did I know when researching "Bigfoot", you would encounter so many profound streams of information – from reading about not only our information on earth but also witnessing some of the most "unusual phenomenon" in our local area. Therefore, I felt it necessary to share this information so the next generation of "Bigfooters" will not have to begin from the very beginning and the newbies have a heads up to what is "ahead of them". The developments that transpired from April through August of 2017 clearly indicated to me that "Bigfoot" has an origin from the "stars". As I go through this research, please keep in mind that I was as perplexed as anyone else – till things just could not be "denied" of the connection between "Bigfoot", E.T.'s., and visitors.

I want to give special thanks to my family for all their support and especially to my sister, Rita Young, for her countless hours in helping me get through the manuscripts and her time reading and proofing all my stories, photographs, and documents. She is an amazing person and I love her very much.

Chapter 1 – Started "Bigfoot Madness" In 2016

Bigfoot Madness was started to research, document, and photograph all evidence in the local area of Carrollton, Kentucky. I wanted to prove their existence but also delve into how they live, document as much as possible, and historically preserve a record of this colony in our local area; this colony has approximately twelve individuals living in the initial habitat site. Since then, I have been researching another site no far from the first one in October, 2017. Almost every researcher that is serious also speaks often about protecting them and their natural habitat they live in. As an investigator/researcher, I try to not leave any avenue undone; this year, I have been in contact with a psychic medium in order to try to get a fix on a language – or means of some type of communication with them.

In each chapter, you will receive a new story and/or pictures; and hopefully, this book will share new information with regards to how they live, new data with regards to their cloaking abilities, how I believe they got here, new expeditions, and surprises with regards to their behaviors.

I will try to present something new, interesting, and visual to study (the engineering and nest building capabilities, noises/social behaviors of my interactions with them, photos showing how they group and hunt, footprints, as well as other finds from the woods). This book will surprise you at times; it may even make you "have to consider another possibility with regards to where they came from". My next investigation/expeditions (after completion of this book and publication) is to delve even harder into the "communication" part of being able to possibly ask them questions and hopefully with my friend, Barbara Ellen McKittrick, will try to utilize her talents in a telepathic type of communication "speaking with them". I entered this area after many researchers revealed some of them have the ability to speak telepathically with them. I met Barbara Ellen McKittrick, author of "Yearning To Speak To Heaven" at the First Annual Bigfoot Expo held in my local area in September of 2017. I felt she would be an ideal person to accompany me on some of my expeditions in hopes of getting a "channel of communication with them".
We all are interested to "know more" and "also what is their purpose for being here with us on this planet". Barbara Ellen McKittrick (also has been abducted in a UFO situation) and recalls some things in regards to being taken. She advised me, however, she has also asked that she "not remember some things with them taking her". Her special talents are very much appreciated. I have requested that she accompany me soon to get closer to the UFO – that I have been photographing this summer of 2017. Although, I usually am not frightened; I feel a little more assured having someone with me on this expedition. As I am not a degreed scientist, you do have to protect yourself. What radiation or other unseen things as we get closer to it – may happen. We will find out; we are planning to enter this realm of discovery before summer's end here shortly.

After my first sighting in March, 2013 of a Bigfoot crossing a highway in front of my jeep, I just could not leave the subject matter alone and had to start working on getting a game trail camera out to see if I could record anything. I also went to lectures, read profusely every article I could get my hands on, and listen to other researchers and their experiences. It paid off. In October, 2014 – I got the first picture of a baby (in front – very low on the ground). He has a very wide forehead, totally black skinned face, hair over his body, and is sitting chewing some leaves on a stick. No far off (left side – in the background), you can see a large figure leaning and looking forward. Anytime a small one is threatened; I do know the bigger ones will come. In this same area, I would leave a treat for them and try to barter. I do recall one time knocking and out of nowhere, it banged (four solid bangs in a row – very rapid). I knew this was an indication that he was sending a message to me and the people I was with. This is <u>my territory</u> and I am in charge. We left immediately and headed out of the woods.

The photography that I use mainly are game trail cameras. I have been with researchers that use computers, lead wires/lines, and sets out many cameras at night and views them on a screen. I think it is up to the individual to decide which he likes the best.

I use game trail cameras as I feel it is less intrusive for them. I think a lot of wires going out in all directions; that they know they are there. You have lots of lead wires and connections. It is felt for my own personal use a game trail cam is just one small spot in their home of the woods. The woods is their home and a private place of security. I first started with the flash game trail cams and soon learned that the non-glow is much better. They do not like any bright lights and flashing.

I also try to disguise my camera from them as well as any human that may wander into the habitat site. I have had cameras taken and they are a hot commodity for anyone interested. Most of the time, I pick an area that I know to have some kind of "true activity" and "also where I know other humans will not go or hike through there".

In the past several years, I have a system that I normally use. I try to always go in a calm peaceful frame of mind. I do say a prayer for them and also for my safety before going into the woods. They like treats and they are always left for them in about the same place all the time.

(Photograph #1 – Taken 10/30/2014-Carroll Co., KY)

Getting this first photo was so encouraging to me. I just could not give up and needed to know more about them. Then, I stepped my game up and kept on pushing to see just what else was out there and how much of it "that I could get a photograph of".

There are many, many stories of the "Bigfoot Colonies" in our local area of Carroll County, Kentucky. However, over hundred years ago, they talked even less about them. They are speaking up more today; but it is still a struggle for some to understand that they are there. I think the general public envision them as a sole "flesh and blood" creature like a deer. You can track and see them all the time. This is very untrue. This creature has capabilities that our other creatures of the woods do not. He is smart, fast, and has been bred with capabilities not known to us. I have learned of these special capabilities and have witnessed them first hand. I will share that they do have the ability to cloak, weightlessness, and zapping. The energy that they have (I cannot determine if it is an automatic thing that happens or if it a controlled act they know they can draw upon and activate this activity). It is one of the best defense mechanisms they have and could be used in their hunting as well.

"Chapter 2" – The Colony

I have learned from my last book, "Kentucky Squatch'n" that the general public is more interested visually in our evidence than a lot of storytelling. So, this book has been put together with this in mind. I want to hold the interest of and give as much "credible" information and truthfulness with regards to my stories. I do not embellish any incident nor do I tamper with any picture. The only thing that I have done to the SD cards taken from game trail cameras is to enlarge them so they will print in this book more clearly for your review and study. I hope you enjoy the pictures of the colony.

It is a portrait of "one of a kind" I believe. The father is going through the woods and right behind him is a young juvenile and behind the juvenile is the "mother" I believe. I want you to take note of how big, strong, and powerful the fists/hands can be and are used not only to defend himself but his family. I have witnessed first- hand one evening a "young male" swinging these fists and using them on something – either to subdue another male and teach him that he was the boss in that colony or he was absolutely pulverizing a deer to eat. I could not see the ground but he was swinging his fists "right", "left", "right", "left" and hitting the object with great power on the ground. All of a sudden a hand signal (no noise could I hear was made) was given by another in the background, he looked up and saw me and hopped up the small hill and out of my sight in the darkness. I could only see he had a very, dark black waxy face to him was all. I stayed put and decided it was not a good thing to pursue. As I was hidden and safe – "no sound verbally came from the second one to tip off this guy". However, the signal must have been clear – he abruptly looked straight towards me, hopped up and left the area. I felt it was as if a hand signal? Could not tell entirely. In the next photograph, you will see the huge muscles of his forearms, legs, and in his chest area. You can even notice the skin on the chest and abdomen of the male. He has two prominent areas where the skin is visible. The arms come down below the knees and just a little below that (almost halfway between his knee and ankle). He is a magnificent specimen to look at; his strength is apparent and overcomes me (how much stronger they are than we are). Please note his crested skull in the photo. I believe by his prominence; he is the "head man" in charge of this group.

A chimp is about three times stronger than a regular male human. This particular "Big Daddy" male is protecting his family from the creature in the tree. Personally, I think he is leading them to safety -- out and around, away from the creature. I believe the juvenile in the tree (quite honestly is a younger, male testing the older one). As far as the monkey or primate family goes, we do know orangutans suckle their infants up to five or six years and the youngster leaves the mother at around nine years of age. Does the "Bigfoot" mothers do this? How often do they reproduce (once a year or once every five years like some other primates) and not leave the nest till they are about nine or ten years old. I believe since this juvenile is holding onto the arm of the father for protection they do not leave till about nine or ten years old. An infant (oftentimes) has been seen on the shoulders of the mother. Their stature (growth) rate is about the same as us (humans). The mother in the back is protecting from the backside and following along. A lot of sightings of this creature has this formation of order. The "Big Daddy" leading them and the mother and child (following a short distance behind) for protection or to the side. Or, since we do not know all the pecking order of colonies, could the young gray holding onto his arm be another wife or mate? How many mates do they have? I am not totally for certain on this – all, I can tell you (at this point) is that he is very protective of this small group. I believe you can also see two lines of dark hair coming up/slanted from his eyes. He is more the orange/brown type color.

Please keep in mind so very little is known about them and their specific social orders. I try to pick every piece of information out of a picture of them to get a more detailed look at them and their social behaviors and interactions.

The juvenile is quite hairy faced and has short gray hair. Her ears are visible on the top of her head. The hair looks about an inch in length all over her body. The male's height alone must be close to 7' to 8' tall. He would weigh quite heavy (700-800 lbs) and could lift over a few hundred pounds and hurdle it with no difficulty. His hands are massive and used to hunt with, for defense, and also their fingers can pick up very tiny things and open and close things (just as well as we can).

I have also heard in the original habituation site a bear and a "Bigfoot" going at it in the woods. It was about 8:15 a.m. one morning and I wanted to go out and investigate cameras (etc.). This was the first time I ever heard a "Bear" <u>growl</u> and a "Bigfoot" <u>howl</u>. They both were territorial – either over food, their young, or that particular parcel of land. The "Bigfoot" howl outweighed the bear's noise and in less than six seconds the bear high tailed it back out to his original spot. The "Bigfoot" won without a blow practically – the fierceness of this howl overwhelms you. I turned immediately and went to my jeep and got out of there myself until a safer trip later down the road (but this was not the morning to go exploring for me). "I puckered and left" to be honest. I always like for at least two individuals at a time to go to the woods for safety. For safety sake, it is always best to have two people researching together.

11

Photograph #2 – Juvenile Hanging In Tree (With Three Figures Walking Far Right - Bipedal) And Two Babies (Left Side Of Photo)

Please note that they are moving through the woods during daylight on a beautiful day (October 22, 2014 at 12:14 p.m. - in the middle of the day).

*Enlarged photo of the last picture – "The Family Portrait".

I am so very proud to present this picture of evidence. If you look/study their behaviors closely, it indicates emotion, love, protection, guidance, and their family unit is in order indeed.

A game warden identified the "creature" in the tree as a bear. I do not agree with his findings; the reason is this juvenile is leaping frontwards out of that tree like a primate. Bears normally face the tree and will slide back down.

The gray juvenile appears to be around six to eight years old. I do not think it is old enough to fight off intruders or bears yet and therefore, the "protective parents" are nearby and look out for their family as we do our own. The fingers are nimble and can pick up small objects and work just like our hands (open doors, make tools out of sticks, bones, and rocks), and walk upright just as we do. I have seen the young ones run on all fours as in a gallop (like a horse) front part of limbs and second pair of limbs – legs coming forward just as a horse would gallop. The massive hands can be used for removing the hide from a deer or other animals they eat. We often find fur pulled out and left.

This gives them extraordinary speed in running down prey for food. The "Bigfoot" can consume almost anything we do (deer, turkey, turtles, fish, wild nuts and berries, drink water from the springs, and birds, etc.) <u>only</u> they kill with their hands and can also use the wood pieces they pick up and knock with – to knock out their prey to eat. They can run down a deer – disable them by breaking their legs in the back and twisting their neck. Their long, strong fingers and hands (remove the hide from the deer by skinning it) and eating the parts of the deer they want. Others have reported vomit with fur in it. They can also pick up a large animal and slam it against a tree (such as a deer or calf). They also use dead animals hit by cars on the road for a food source as well.

Some researchers have investigated boneyards left behind by a colony of them. They have found dogs, cats, pigs, horses, and calves' skeletons piled up on the ground that had been consumed. They do not kill more than they can eat and only harvest what is necessary. I have not found a massive boneyard yet; but I can still search for this – I want to see it for myself. I have seen single kills regularly though. Sometimes, I find a leg ripped off and the rest of the body missing. No normal hunter in the woods would do this act.

I spoke with a lady from Sulphur, Kentucky who reported seeing one huge "Bigfoot" carrying two deer (one under each arm) down the road and out of sight. If he was selfish or super hungry – he most likely would have killed one, sat down, skinned it, gutted it, and ate it. However, he was carrying two grown animals away – intact. Was he taking them back to the colony to feed his family and little ones – provide protein for them to survive? This is not a selfish act of any kind and one that relays they have a lot of human characteristics to them – even though we do not have all the pieces to the puzzle yet. We do find, however, <u>each picture tells a story in itself</u>. This story certainly did for me. Also, look at the bottom left side of the picture (just presented to you), you will also note two small individuals sitting on the ground.

The one looks (in the face) like a baboon. His color is an orange/brown and he is sitting exactly like a primate would. The one next to him on the right is completely black. He also is sitting in a primate state. The baboon looking one – has a longer snout (or nose area). I have seen them flat faced, protruding nose a little and a lot, brown with a beige colored chimp looking face exactly, and also a very hairy face. Could this be due to inbreeding or is that just the way they look – diverse just like us as humans are? Not all our babies has small noses – some have rather large noses and lots of hair while others have no hair at all. After my UFO research this summer, I think I may have a partial answer as to why each one looks so different.

Also, some have been real hairy all over their bodies as others' hair is quite light colored (gray, or coyote coloring – gray/silver) and short. We find their characteristics intriguing as they always know what "color" to hide in front of to not be seen. When a light colored gray one is in front of a light colored tree – you cannot see them. Same way with a dark brown or black, they always stand in front of something dark or in between two trees to stay hidden in the shadows as they blend in. It is amazing how they know to do this. Occasionally, you will find a very young one that is seen, because he is not fully grown, <u>immature, and playful</u>. I think this is what has allowed us to see the babies more. This happens especially when they are frightened – they let their <u>guard down</u> a bit. I saw one about 2 p.m. in the afternoon sitting in a field; all black in color, squatting. This is very unusual indeed. He looked like a stump until he moved. Nobody would have noticed his presence at all that day at that time. I picked it up by chance driving by. I thought there was no tree there – then it moved.

They do build nests on the ground and in the trees for shelter. At night, I have a picture of one baby covering himself with leaves and sticks to keep warm, stay hidden, and they do huddle and play and sleep together. This would give them social interaction, love, warmth, and "family unity" for sure. Just as our own kids bunk in the middle of the floor in sleeping bags – they bunk and play together under the trees at night and in the day. They sun themselves, groom bugs off of each other I am sure – like any other primate, and we often find photos of them laying on the ground six to seven at a time piled up in a heap. This expresses love, unity, and family. Some researchers have reported (10 – 12) individuals in their research areas or colonies. I have counted approximately (12) in the first habitat site set up. I am in another habitat site now and continue to count and research this second area in Carroll County, Kentucky. It is too early yet to determine how many are living in the second colony found in our area. I do know I have taken a 15 ½" print out and took a picture of a 19" print over the past couple of years in this second site.

As a researcher, I cannot get enough of the story about "Bigfoot". It comes in tiny pieces and takes <u>lots of patience</u> to keep putting little tiny bits of information together to come up with a "whole story" but I keep trying and am not giving up on this. There is way "<u>more to learn</u>" – what about how they get natural medicines. Is this from leaves, plants, rocks – I saw a monkey in Brazil take a rock and hammer it on another and then try to ingest some rock dust – which would have properties of healing possibly – this is interesting – does "Bigfoot" not only use rocks for communicating (clacking), hammering and making tools, used to cut food or possibly tenderize it, but now is there a medicinal possibility as well. This was very fortunate to see this piece as I have learned a great deal about primates' behaviors in order to see "how much" that parallels our own "Bigfoot" in Kentucky.

I am so very proud to live in Kentucky and be able to see this creature (close and from afar) in order to study and photograph it. You don't have to travel far in Kentucky to see one – go to the woods. They must have woods to survive and "we must always protect their environment as much as possible". Orangutans are being rescued for this very reason today – woods being cut away too rapidly and the animals have no place to go to survive. Orphans have been pulled and ripped from their mothers' arms and must be trained to live in the wild again after rescue. If you want to see a great show, watch "Orangutan Island". It is a phenomenal show in every way and what a worthy cause to save these creatures and study them further.

We, too, want to save our "Bigfoot" in Kentucky for further study.

Chapter 3 – Unexplainable Things 16

Occasionally, as a researcher, you will come across some kind of unusual, and in fact, unexplainable situations and features from the photographs you gather and study. All of my pictures are taken with game trail cameras and have served very well in trying to piece together information about Bigfoot. However, I have had unusual things show up – and therefore, I have created an "unknown" small chapter here. I can only give you my theories as to what I am seeing and you can come to your own conclusions as well. There is no black and white on these issues – and as humans, we are still discovering, researching, and trying to make sense of it all. This chapter may contain unidentified creatures, orbs, strange ghostly figures, energy photographs, etc. As a researcher, I feel they should be documented here as this is often seen in the field and experienced. Even though we cannot explain things (due to our lack of abilities) does not oftentimes mean it was not there. Hopefully, I can get some good comments back on these to shed further light on these topics. I am curious if anyone else has seen this *"unidentified creature* in Carroll County, Kentucky area".

In November, 2014 – please note the time / date clocked on the game trail camera, it picked up three very unusual pictures. There is an extremely large bird sitting on a log. It gets up, walks bipedal, and then runs away on two legs. The head is white, gray long beak, white/rust covered body with either feathers or fur, and gray legs. Due to the location of my camera, it overshot approx. at least 30-40 feet from the norm and picked this up. "Could this be something unidentified totally or something thought to be extinct that did not go away?" I have estimated this thing to be at least 6' to 7' feet tall (due to the location of my camera and knowing the surrounding area).

Photograph #1 – Bird Standing Bipedal (Running Off – Mouth Open)

Figure was frightened by something–was photographed running off on 11/1/2014 – 6:02 p.m.

Photograph #2 – Large Bird (Still Standing)

Figure standing on 11/2/2014 – 8:47 a.m.

Photograph #3 – Bird Sitting On Log

Figure sitting on log 11/3/2014 – 10:00 a.m.

I will share my theory with you about cryptids later. These photos were approximately two years old when I started receiving some more information which would change my thinking about a lot of things, especially how and where so many types of cryptids may come from.

Chapter 4 – Evidence

Under this section, you may see such things as footprints, photos, huts, nests, and an array of other things we find (as the evidence comes in). This can also include additional stories from the past and at present. I hope you enjoy this feature of the "Bigfoot Madness" book.

Now, you will view about six (6) photos taken from 1/30/2016 to 2/7/2016. Something to remember about this story is that I look for places that I know there is not a lot of barefoot traffic and also (since we had a snow) – by February 7th, the ground was a little moist and you can easily track prints in the mud. I called a Bigfooter from Madison, Indiana and we took out together on this trip in Carroll County, Kentucky. We found two sink holes which is a great place to look for prints. We noticed a trackway immediately and got busy casting. We came up with several good prints to show you and also a hut built.

That day we found two young juvenile prints and several baby prints. The prints measured approx.. 5" wide – 9 ½" long, 5 ½" wide to 10 ½" long, and three sets of baby prints 3 ¾" to 6" long and under. Lots of times, I will give a friend a set of prints as well as the property owner for giving me permission to go on their land (as a keepsake).

Also, you will observe some of their hut building. One that I like is the way one smart one, used a branch to come out and support his hut – just as we would lean a branch and then fill it in with vines, leaves, and other items from the forest floor to make it more warm and weather resistant.

I often find them utilizing fencing to tear down, roll up, and make igloo structures to fill in with leaves and brush (insulating themselves from wind, rain, and the elements). One important thing to remember is when you hike (look for trees pushed over fencing). This makes for a bridge for them cross – especially if there is barbed wire on top the fence. I found this as one site I went to – there was just too many trees *pushed over the fence – they all didn't fall that way.*

Notice how he has used one branch to hold up the end branch on the left and then he filled it in and "nobody" would even think this was a nest (probably most would just think it was a bunch of fallen branches in the woods).

Picture of first juvenile measurement.

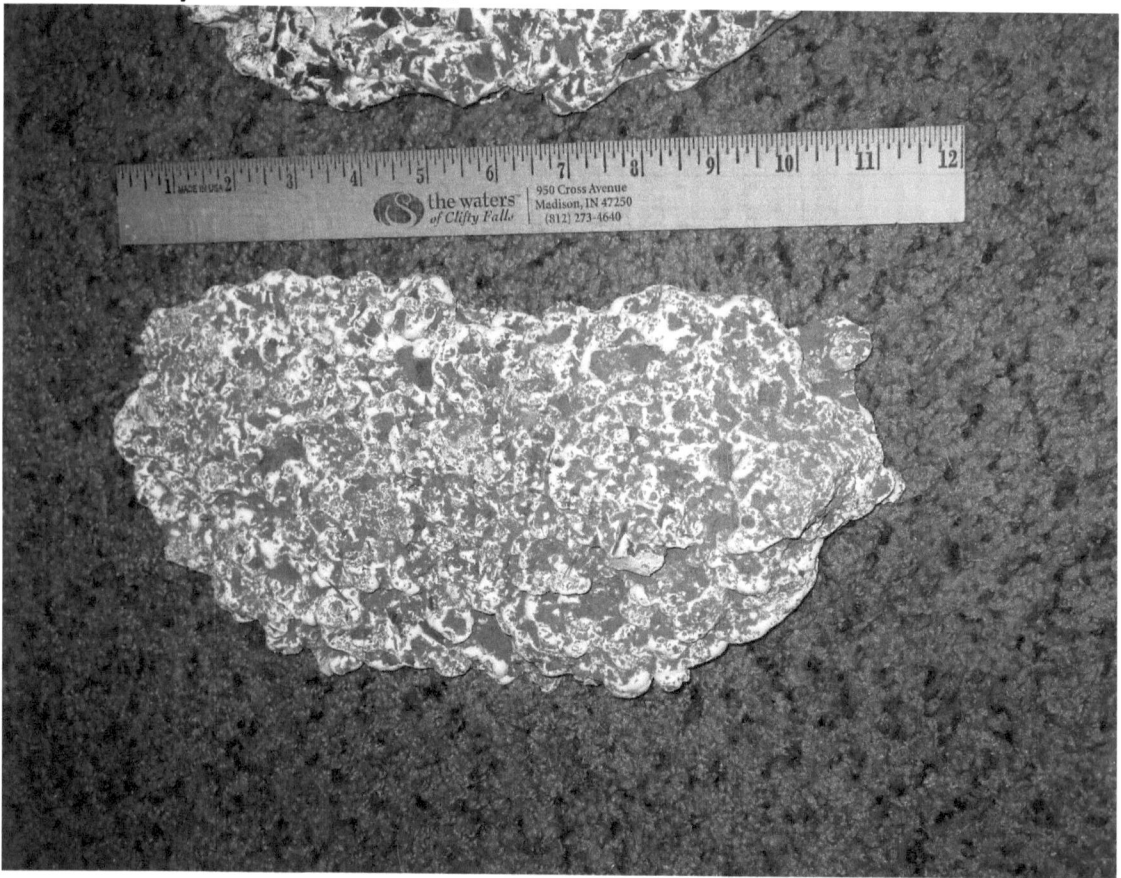

Nice photo of four sets of prints of varying sizes (juvenile and babies).

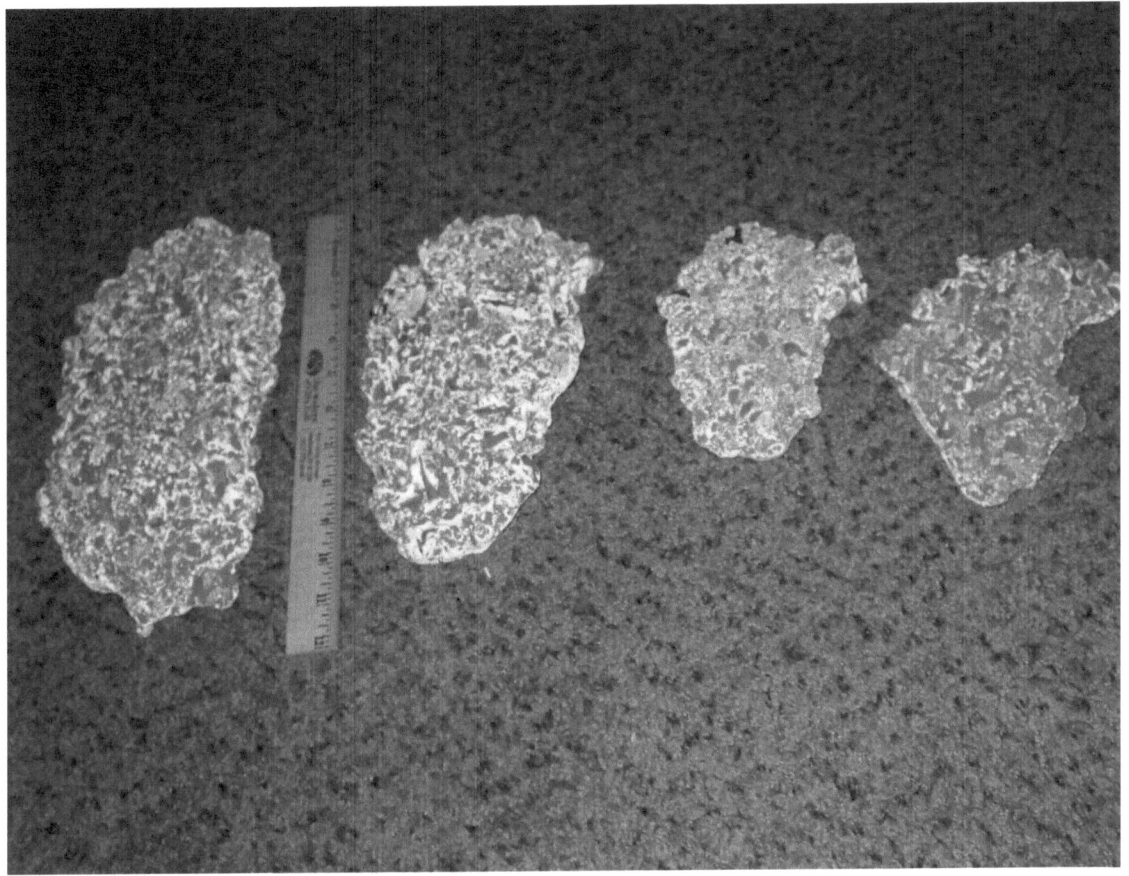

What I like about this set of "baby prints" is the fact the one on the right (when he stepped down), he flexed his foot and it turned slightly – which we cannot do (as our foot is just not made that way). They have that middle break which gives them great flexibility in climbing/grasping trees with their feet as well as hands.

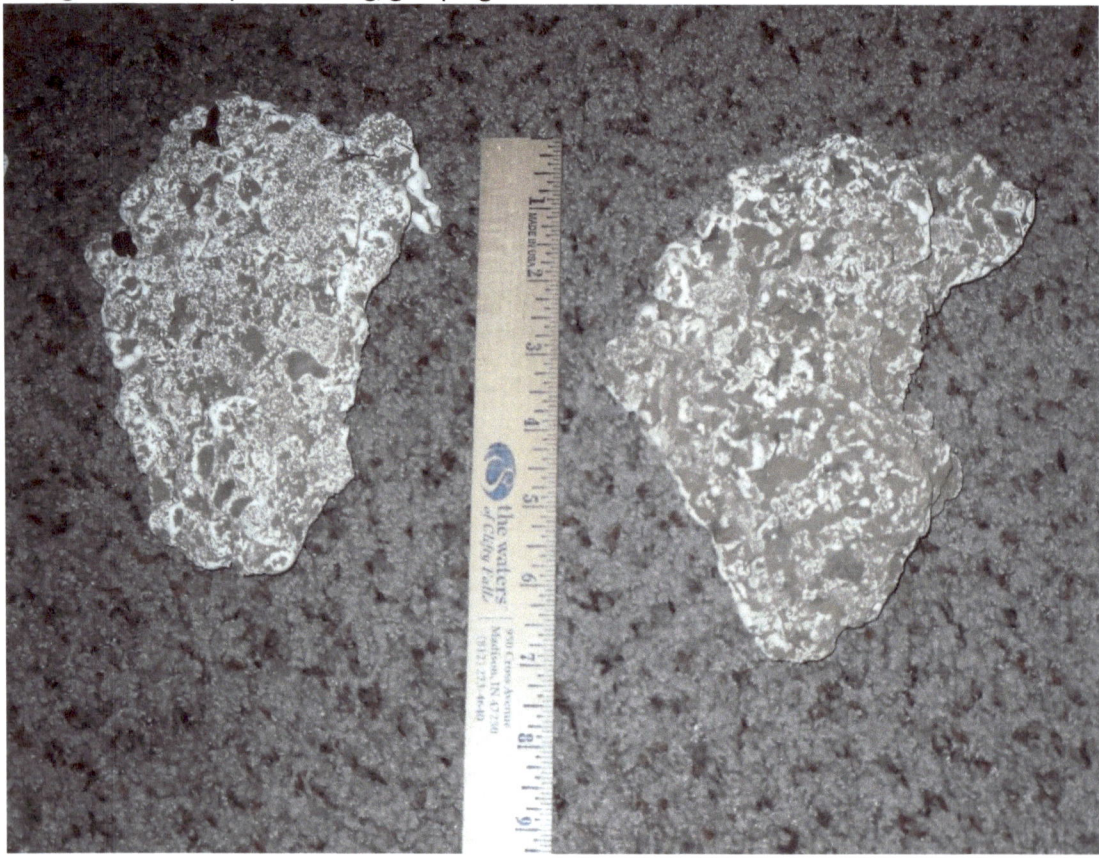

This gives you a different perspective of what they looked like before we dug up the "baby prints".

This photo just was so unusual – a circle of materials – just did not look naturally made by weather and forest growth. "I definitely think this was put together, a perfect circle of materials." Next question is "why"? That is yet to be known.

Chapter 5 - Photography

I normally do more still photos but am honing my "video" skills as well. As a result, till I can get a lot of blur out of them - - I do not concentrate too much on videos. So, I will insert another interesting story for you, and I hopefully can get a lot better at this to record expeditions better. As a researcher, I have learned you have to be very patient and *let them come to you*. I do not really chase them at night – you never know what you are going to run into (coyotes, bears, pigs, wild dogs, etc.). It is much better to rely on what is comfortable for you and building a small fire and sitting and relaxing (once the fire goes out) - - there will be quite a bit come into check you out. I work many, many hours reviewing photos (in sequence) to see what story is being presented – how is the photo in front of this one different. If you view your materials in this manner, it will tell you lots of interesting information as to what they are doing in their habitat. All of my photography can withstand any kind of testing and scrutiny – no trickery; I guarantee it.

One of the experiments we tried in the original habitat site of Carroll County, Kentucky was to put a birdhouse in the site. I really wanted to see what the babies would do with this. I had been gifting them with treats and would leave. This is the results from this activity. They do recognize you and get used to you.

After I left a gift, I noticed things were beginning to be deposited in the bottom of the birdhouse. As a result, I asked another "Bigfooter" with me to come join me and bring the birdhouse in. We wanted to crack it open and see what he was leaving. I did observe one evening about 11:00 p.m. (after one deposit) – he displayed the following behavior. He had hammered for about 10-15 minutes and then "hooted and clapped his hands". This was, I felt, as if to say he was proud of this accomplishment but also he had received his favorite food treat from me as a gift. So, as a result, we brought the birdhouse in to see "what he had left". This kind of gave us an indication of what he thought was important, what kind of gift he left us, and how he felt. I think "he was proud" to have left these "treasures for us". I was glad to get them. I think the "hooting" behavior was because he found food. I think the "clapping behavior" was to say he had accomplished hammering through the holes the "gifts" he wanted to leave.

This activity he displayed told me he was "happy", "industrious", and a "very busy little guy" in the habitat. He had been depositing things all along. I will tell you though now I wish I had not opened it in a way. After it was cleaned out, he did not leave any more materials he found valuable. Wonder why? I was still bartering with him.

Photograph #1 - Picture Of Birdhouse Brought Inside (Before We Broke It Open)

Maggie, my cat was assisting and being quite inquisitive to our tearing this thing apart to see what was inside.

After the top was taken off, we found a child's screwdriver, two wooden knobs, a paper wrapper from gum, an old roll of used film, and a baby doll's hair brush.

Photograph #2 - (The items laid out on a towel to photograph."

Photogragh #3 – A researcher looking at one of the wood knobs he left.
  *Note: The knobs had been ripped off a building or outside shed with his fingers.

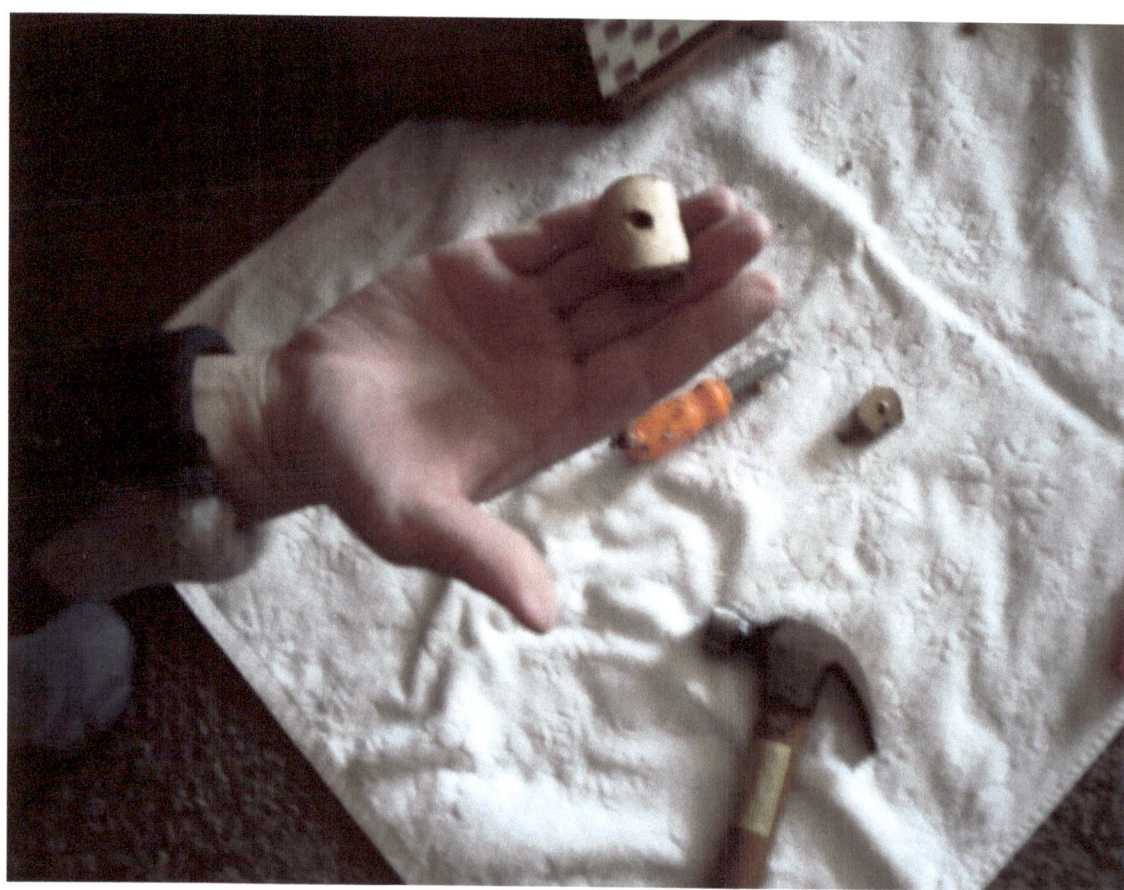

The strength of tearing these off a building or outside shed with his small, tiny fingers – wow….
As I said, their strength (even though they look cute), could not be a match for a regular person even though they are small. I viewed a gentleman trying to catch one once. I thought, "What is he going to do with it even if he could?". He was no match; he got away. The treasures he left was probably gathered as he made his little rounds at night. They do go through housing areas and are not afraid to occasionally "scratch" a window screen or "peep in". They are curious and playful when they are little. Their little tiny hands/fingers can pick up anything and carry it away. Babies are inquisitive and crossing yards at night – no problem for them.

Chapter 6 – My Baby Pictures   *31*

As I begin to continue reviewing pictures, it is amazing how much data, pictures, and stories have built up since the publication of my first book, "Kentucky Squatch'n". I continue to go "Bigfooting" alone and with friends occasionally and find it to be a very enjoyable hobby. It is interesting, relaxing, educational, and helps to get me out in the woods, and enjoy so many other "Bigfooters" in our area.

First Photo – Taken December 29th, 2014 – 2:09 p.m. As you notice, he is out in the middle of the day – but well hidden (low, dark place, bottom of trees).

This subject is using his (right hand-forearm) to pick up a small club on the ground. They can use clubs to hunt, build, play with, and communicate by knocking. I have heard them knock with a club for an hour at a time in the original habitat site (as if building something). They can make elaborate huts with the sticks as well. This baby has features that resembles a small "baby gorilla" – hairy all over and brown. I find sticks stuck randomly in the ground and propped against logs (for digging grubs).

Please notice how well he grasps with his hand. He stays low, close to the ground and hidden behind a small sapling. They often do this to blend in and not be seen. They always put something between you and the camera almost always (this is a protective behavior). They do know exactly what color tree that matches their coat of color to blend in. Each photo that is included in this book is being gone over thoroughly so as not to embellish any data and only present the facts.

If you like researching, "Bigfooting" is an excellent hobby. Do not get discouraged though as it takes time researching, putting out cameras, and going back to research the photographs that came through on your game trail cameras. This can take several hours and days of looking at evidence in each picture to be sure "you did not miss" anything in the photo. I try to look for a story being told in each photo:

1.) What were they doing?
2.) Were there any other animals present and their reactions to them and their surroundings?
3.) What does each one's appearance, stance, and facial expression might "explain to us".
4.) What can you see from this one instance that is "happening"?
5.) Can you see anything new or "unique"?

Evidence – 1/2017

This small baby was photographed by a trail camera on December 15, 2014 at approximately 2:47 p.m. in the afternoon. He is staying hid "just as he has been taught" – low by the dark part of the tree – in a primate stance (sitting). His muzzle is a "light beige" and his hair is dark black, about 1" all over the rest of his body. It appears he is observing something straight ahead of him. I figure him to be approximately 3' – 4' tall approximately (if he was totally stretched out). He looks to be about a year or a year and a half old possibly. I have some pictures showing small ones on the ground and alone. You can bet the "protective parent" may not be far off from them. How soon does the mother put them on the ground? I believe as soon as they can scamper some on their own. Please remember, even though this looks like a cute, cuddly chimp – they are extremely fast and strong. I do not try to get real close to them "as I am not sure how aggressive they may get when frightened". I have had one of them try to play with me by "throwing a small pebble at me" about 10' away. I did not pick up a "pebble" and throw back as – *could this be viewed by it <u>as play or aggression</u>*". And, at the time, I did not know where or how far off the protective mother might be. It would be best not to chance it.

Photograph #2 – Taken December 15th, 2014 – 2:52 p.m.

This picture shows the same "baby" in the same place. He has moved just a little bit but not much. He stayed put about five minutes to be photographed by the game trail camera. I have heard babies this summer give out a long, whistle to let a parent know there is an "intruder" in the area - <u>a long warning whistle</u>.

Chapter 7 – Bears

(Just had to show you the picture of bears in Carroll County, Kentucky). I would never have dreamed they are migrating over to us and producing very quickly.

As I continue to research in the original habitat in Carroll County, Kentucky, I had a very unusual picture come up of this "bear" with her cub. The excitement of finding one in our territory was extra special. I had been told for a long time there were no bears in our county for a very long time. Of course, this was bad information from a hunter or two. I am happy to capture them on film; however, why are they not hibernating at this time of year by now? For this reason alone, I have got my concealed weapons permit and do carry a revolver on me to shoot in the air (hoping to get enough time to get away from a bear if possible). The little cub is resting so peacefully by the back flank of her hip and seems "mother" is totally exhausted. A mother can get as large as 150 lbs.

Chapter 8 - The Rock Experiment

As I continue to build up my evidence and learn how to do testing much better, I will listen to others as to what has been tried, what succeeds, and what doesn't. Sometimes researchers will put out rocks on a log or something and see what they will do with them. Will Bigfoot take the experiment away or do something else with the rocks?

So, I thought I would try this experiment in Carroll County, Kentucky to see what one would do and get the results. The next picture is of a table put in the habituation site. I placed five (5) rocks in a line, approximately 2" apart from each other. I left a food gift for them and the next morning; I had a hit on the table. Please note when you see the first photo, it told me nothing. It just looked to me like it moved the rocks around. But, then I got to thinking I was looking at the table from my viewpoint, not from the view-point (the way it was standing). Then, I went to the other side of the table to see what it looked like. It looked as though he had drawn a picture of himself, possibly of me – seeing me come and go, but his behavior certainly exhibited intelligence in that he could form this "picture with rocks". I thought it was amazing – and, was that my gift for the day. He had left it for me to see. It was plain; he was creative, communicating in his own way, and to form this and manipulate it in the manner he wanted with those fingers.

Picture #1 – (From My Viewpoint):

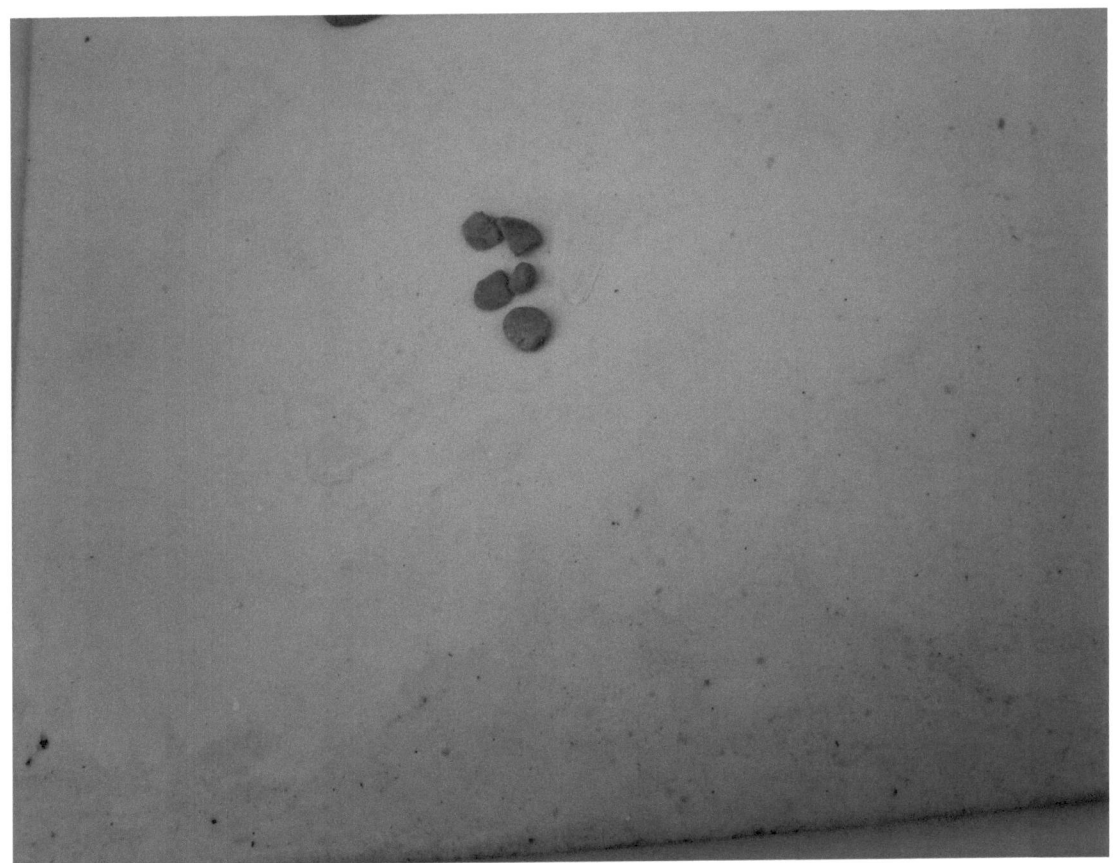

I was positioned on the opposite side (otherwise I was looking at it upside down here). However, when I went around the table and viewed it the way he formed this "art". It then made more sense to me. I stood exactly where "he had to stand" to manipulate this.

Picture #2 (Viewed The Way It Should Be):

*This is the exact picture of how it looked "when I viewed it from the position he placed it on the table". You can see he arranged the rocks in a picture formation – looks like a "stick figure he has drawn with these rocks". Quite a few researchers have tried this experiment and been very surprised by what has been left for them. Was he drawing himself, a playmate of his, or was he drawing "me"? You have to ask yourself; "What was he trying to communicate here?" This took him <u>time</u>, a <u>logical thinking process</u>, and <u>dexterity</u> to place such small objects in this form. This showed "skill" in his behavior and intelligence.

Chapter 9 – The Unknowns Of Research

The "<u>Unknown Section</u>" of this chapter is probably the most difficult to write about and explain. This section will contain information on orbs, paranormal, electromagnetic fields, and other anomalies found in our "Bigfoot Research in Kentucky". They are also the hardest to prove as well. I will describe to you "how I feel about the photo" and what was happening and how "I felt at the time". Please remember, each person may perceive this data a different way. I will address at times portals, mysterious lights, energy, and other "true experiences" I have had in future chapters to come. Many researchers are finding very unusual things in the field that cannot always be explained; but we do feel that we need to <u>document</u> these occurrences for historical records (for future generations of those interested seriously in this topic). In the past, lots of information was lost, misplaced, not documented, and the history of the "find" was lost forever. Long bones and other artifacts treated badly and lots of the evidence turned to dust – never to be retrieved again. This could have a great impact on piecing the "story of Bigfoot" and his ancient history together.

In April and May, 2016 – I had three (3) unusual photos appear. I hope you enjoy the photographs; you will see energy and light exhibited, an orb, and also a peculiar looking face in one of the photos. I will use arrows to point out "what I am seeing". Lots of researchers are seeing unusual things in the areas we research like this. Even though some of them are not willing to discuss these activities – I do believe there is a connection somehow between this creature and these activities. May we learn more about how there could be a possible connection in our research as we go along.

Photo #1 – Displays A Mist – Not Fog

This is taken in Carroll County – Research Area. The photo in front of and behind this – did not display anything like fog that night. We have cast several prints in this area in the past of "Bigfoot".

Photograph #2 – This photograph displays an orb, strobe of light, and you see something like "eyes" bright and to the right of the tree.

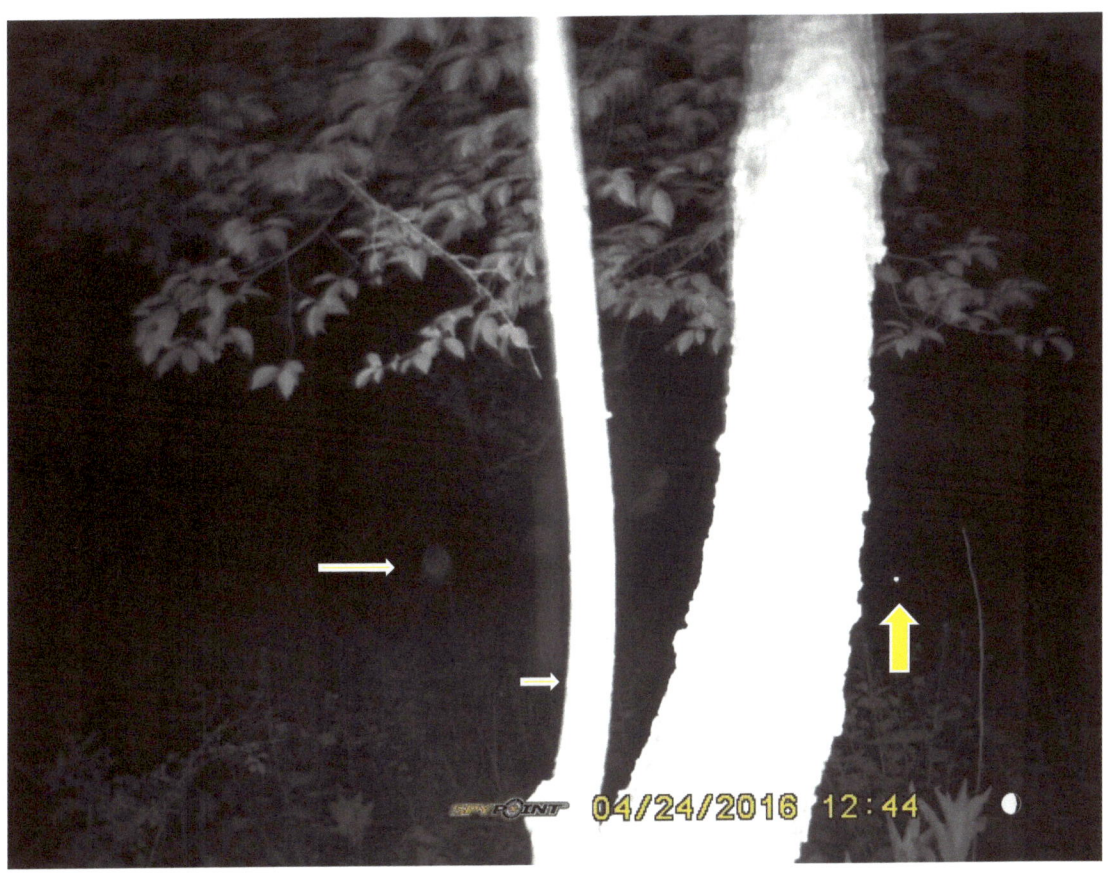

Photograph #3 – You will see a face at the very edge of the photo (left side) with a very light colored outline of a body in the photograph. This was taken in the same area of the small baby prints, orbs, and mist. I will put one arrow facing down towards the <u>face</u> and one arrow pointing <u>out</u> the light outline of the body of this figure. Also, this is a very low, lying (sink hole) that I found the initial footprints in. This figure is reclining; he is looking at the game trail camera. I see ears on the left side of the head and very strong, facial bones – no hair on face. This creature positioned himself in the manner a "Bigfoot" would. He put the tree between himself and my cam. I got a partial photo of the face/body. His body is muscular and curved (leaning to the left). The reason his body is light in this photo – he truly is trying to do what I call "<u>cloaking</u>". This ability to conceal and disappear (at will). My camera must have caught him halfway through this process.
 *Notice the waxy looking face (first arrow on left-right below the leaf).

Chapter 10 – Powder Appears

This picture must have been my "Christmas" present, it is so fantastic to me. I call this "baby – Powder". His face is extremely white colored, his forehead is very thick – and you can see a powerful thick bone structure there, as well as a very wide nose. His mouth is pretty much like ours and his head is very round shaped. He is also hiding in a dark spot observing his surroundings and what is in it. I like this one so much as it looked very similar to one I saw on T.V. in Canada. His eyes are very round, indeed. It reminds me almost of "Neanderthal" looking. Knowing this area very well, it is also approx. 3 ½' tall.
<u>Powder (12/26/2014 – 11:25 a.m.):</u>

Same Picture Of Powder: Will try to enlarge for better view of his facial features. He also has wispy hair on top, his face is a powder – white. His eyes are very, very round and there is no hair on his face (some on the body). His lips are very prominent, white.

Chapter 11 - September 18, 2015 Expedition

As a lot of our expeditions are done in Carroll as well as Trimble County, we will often hook up with friends and go out to share information for a night or two. Most of the time, we like about four to five people in the expedition and that they be quite dependable – as you never know what might come into the camp at night. It could be a "Bigfoot" or it could be a "bear" or it could be a "coyote". You want to be sure someone has your back and is an expert at "fast <u>getaways</u> if need be".

This story takes place in Trimble County, Kentucky. We had approximately six people in the group. All of us had some type of "Bigfoot'n" experience. This evening and the next night would prove to be so informative in many ways. We all had experiences together; we all had individual things happen to us we cannot explain. I will not name "Names" in this group as to protect the people involved. I feel they would not like to be revealed.

On Friday afternoon, I met one of the guys and we began to set up the mess / kitchen area and he was working on setting up electrical equipment and cameras for Friday and Saturday night. We were by ourselves and out of nowhere comes a howl from the West side of this property. We were astonished this happened so quick and us by ourselves. We could not wait for the others to get there and get settled in squatching for the night. Later, they filtered in from far and near and all of us were there (all the researchers).

After breakfast the next morning, we all went hiking together looking for prints, hair samples, stick breaks, listening for knocks or howls. Nothing really special happened during the afternoon.

However, about 10:00 p.m. Friday night, I went to my tent to change my t-shirt as it was hot and I wanted fresh clothing on. My tent flap was not all the way down in the back of the tent; but it was facing a sink hole. I hear a taunting laugh, "ha, ha, ha" and "ha, ha, ha" kind of a mischievous laughter – taunting. I did not say anything to the rest of the crew there. I thought it was someone playing a trick. Later, they advised they did not hear it.

Then, on Saturday night, the guys had something weird and freaky happen to them that I did not hear. Later, after we heard grunts from this sink hole on Saturday night for about two hours. At about 11:00 p.m., whatever it was "Bigfoot" or something else - - did not want us here. We hear "grunting" from 11 o'clock till about 1:00 a.m. Then, the two gentlemen in the group heard women screaming and like metal "forging" like from older times. However, please note - - I was only about 15' from where they were and did not hear it (what was happening to them at all).

Also, when I picked up my game trail camera – we had some weird pictures occur. I will show you some of them. They look like a wolf's head with laser beams coming out of it's eyes, nose, and mouth.

What I could not understand when we spoke to each other later was the following and still do not know to this day what caused it. They did not hear my taunting laugh on Friday night and I did not hear their women screaming or metal forging. Then, when we spoke of this story down the road – I asked them if one of them walked behind my tent that Friday night as I heard twelve steps go (back and forth) as if someone got up to check the perimeter since we had so much grunting going on. You never know if something might "charge in". In this case, thank goodness – nothing did – but we were all ears (every one of us – very attentive).

Here is the answer, <u>not one of the guys got up to walk the perimeter</u> of the camp (with their revolvers) just in case. They did not see anyone walk behind my tent. It was just as plain as day – I could hear the footsteps behind my tent (back and forth). Since then, I think I have figured out how it was done and will explain. I think one of the "Bigfoot Cloaked" themselves and came in to check it out. That is how I heard the footsteps.

This was one of the most "special trips" I have taken. We all had experiences, gathered a lot of information – to decipher, and enjoyed the pleasure of each other's company. On Saturday evening, one of the girl's has a very feminine "howl" and out of nowhere a big male answered her back from the "East Side" of the property quickly. Absolutely fantastic to hear this in the wild. We were not able to capture it on tape but loved it the same.

This trip precipitated me; however, beginning to push me to go it alone in my research as an independent researcher. I feel we are going to have to look at a lot of different types of evidence in order to fully understand the creature, "Bigfoot", which also includes paranormal activity, portals, EMF (electro-magnetic fields), etc. At about the time this story was happening, I had two experiences which is turning my research into several different directions to try and solve some of their "mysteries" that we have not understood till now.

I can report what I have witnessed to you and show many pictures. However, it is up to you to decipher what you want to "believe". I will tell you two things: 1.) <u>None</u> of the pictures or stories are non-truthful – they are the truth and no picture has ever been doctored, and 2.) As a researcher, you do not want to dismiss evidence (because you or someone you tell may not believe you). You report the "truth as it happens" and can tell what you think happened and why. However, if you do not know – you tell the public you do not know - - till you can pin it down with a response. I have always faired well reporting truthfully, the stories of "Bigfooting".

Now, I want to show you some of the pictures we took in this area of Trimble County of the wolf's head and laser beams. The odd paranormal stuff – I cannot explain at this time but I think I may have an explanation for footsteps heard; but not seen.

Just about right before this trip, I had an experience in August of 2015 close to the Carroll County Habitat (that was set up). As I was about ready to enter the habitat area early one morning, I felt something was "watching me". It was a juvenile – about 5 ½' to 6' feet tall, gray, peeking from a tree. He/she spied me and knew I had spied it. He was only 25' or 30' feet from me. I was messing with a lock and gear, bent over. He stepped approximately 10' across to the next tree to hide from me. As he went over there, he de-materialized right in front of me. I could see the leaves behind him and only the outline of him remained by the time he reached the next tree. IT DID NOT MAKE A SOUND. HE DID NOT MAKE A SOUND AS HE MOVED TO THE NEXT TREE (NO STEPPING SOUND AT ALL). This thing was totally weightless.

Could this be "what was stepping behind my tent" and the guys could not see him 10' to 15' from them. They were gathered around the fire. Both of them advised, neither one of them got up from their lawn chairs around the fire. So, who was it behind the tent – a paranormal happening, a "bigfoot that has the ability to de-materialize walking back and forth – checking out the tent and the campsite", or something else we cannot explain yet. I believe they use this mechanism they have quite often – that is why they can disappear on a trail with no footprints. After I saw this, I figure they can come and go and be seen only if they want to – to one's they want to see them. I had a friend that would ask me often (especially when we would find logs hollowed out in the dead of winter with 14" of snow on – how they got in and got out). They had left no tracks of any kind; the only tracks we found was our own. Well, that is the answer. I witnessed this with my own eyes.

Now, the question for me is can they <u>all</u> do this and how old do they have to be to be able to trip this "mechanism". Is it something they are born with or is it something they develop – like skills, language, etc.? And, most important, how do they do this process? I cannot answer that one for sure; but I am still looking and researching.

I know several individuals that have followed tracks for quite a while; and they just stop – vanish. Also, noted is – there is not one sound (a stick break, branches broke or anything when they do this). Most important, where did this ability to "cloak or vanish come from"? Could there be some kind of spiritual connection, some kind of physical energy they are just born with – with no spiritual connection, and also – will we ever know where they truly came from and how their ancestry interconnects with our own. How is it if we came from the same ancestry (anthropologically) and we did not inherit this trait.

Pictures From Game Trail Camera (Wolf Head) or Paranormal:

The next picture shows the wolf head with (laser beams or light coming from it's eyes, mouth, and nose). Please also note (what looks like an apparition of a man sitting on top the mirror that was placed there). His back is curved and the wolf head is still there. As I am more interested in "Bigfooting", I am beginning to wonder "why" we are picking up quite a bit of different paranormal activity. Is there a connection? Does Bigfoot play a part with this – is this where their energy comes from and is something or someone helping them? As you can tell, we have way more questions – but, must include this in my research as it is a part of this. I am going to get an EMF reader soon to start tracking energy levels (when a Bigfoot is present or not), the lands we have this paranormal activity on, and see if it lingers after they are gone or if it is just there while they are present – Bigfoot. The picture below shows the Wolf Head (and ghost - with a curved back): (Could it be a "Dogman"?)

These pictures were taken by game trail camera after our expedition on September 18, 2015. The camera was left and when I reviewed all the photos on that SD card, it just seemed the beam of light coming from the sky was way to organized and much brighter than the one ray of sunshine – which caught my eye. When I focused upward, the head looks like a wolf – but the second photo shows a fog or figure shaped liked a man, sitting on the mirror we had put out to attract the "Bigfoot".

I am not a paranormalist and cannot explain the energy involved in this. I am still trying to find out the actual history of this land. The Trimble area mainly had Shawnee Tribes inhabit this area. I recently spoke with a member of the Choctaw Tribe and he advised that Shawnee were the sole inhabitants, settlers for sure, farmers, and am trying to find out more about "possibly the military involvement here" - - if any that could play a part in this.

Maybe he is not at rest? Maybe he is overseeing (what may have been)? I find it interesting that we are beginning to pick this up. Also, this activity is happening close to sink holes (mainly) – what part does this play in "Bigfooting". I find tracks of "Wood Babies" easily in these holes as they are muddy most of the time and they travel two or three at a time – oftentimes finding many sizes at once of their feet. Also, you may only find one print – a good one.

This trip surely gave all the group a lot to talk about – footsteps, screaming women, metal forging, taunting laughter, and "howls" all in a two day – weekend trip. Wouldn't have missed this for anything. It is amazing when I hear people say there is nothing to do – man, there is a lot of wonderful things yet to be discovered in Kentucky for sure.

Chapter 12 – They Just Vanish   52

This chapter is definitely going to make some individuals feel as though (after reading it) that it is way too far out.   However, I must assure you the facts that I am about to report – I have witnessed personnally and they are extremely unusual.   I report them in hopes that if someone else has seen these unusual activities they can come forward and not be afraid they may be taunted or made fun of in truthfully reporting a find.

The first incident was the tree-to-tree walking. As I just described in the last chapter, the "gray one" walked or kind of floated over to the next tree in about ten steps. It de-materialized – I believe because I had seen it. I really do not think it was afraid but more a matter of "he forgot to turn it on" this <u>activity</u> before I got a glimpse of this being done. I did not react "in a frightening or threatening manner". It went over "smooth, easy, and I could see right through it". This only took about a couple of seconds – timewise. No sounds of his feet moving and no branches broke on the ground. With his weight, I felt I should have heard rustling of ground leaves and some small branches. He was only "truly" interested in observing me; and I did not feel in danger but in more "shock" as he was about 25' to 30' away from me. All I could see was an energy line (left around his *outer edge of his body*).

This fact definitely changes the way I think about them. They definitely are flesh and blood – but, what is this added advantage and where does it come from. However, I can now answer the question, "How do they get in and out without being seen?"   "Is this a biological process solely or is there a spiritual connection with this somehow?" A friend would ask me continuously "How do they get in and out?". **Now, we know that for sure – I am just glad I did not scream**. In July, 2015 – I had an outstanding opportunity to either witness what I would call a "portal" or a "ritual". This started in the very early hours about 2:30 a.m. and lasted approximately 45 minutes. The site was in Carroll County (and is one of my favorites) as it yields the most important information to me for reporting purposes. There was a small group of "Bigfoot" gathered around in a circle. At the same time, they were all "whistling" – and, with no breaks in the whistle, not even to come up for air. This began to annoy some living in the area – and the local police were called. One officer arrived in the patrol car – got out of his car. His next action "was not to approach the field or area of the field that was glowing" and the "whistling continued on solid". Could I have witnessed a portal? Or, was it a "ritual" of theirs and had nothing to do with "transportation" of them – if they do go back and forth (which I do not know). Is it in connection with "how they can energize and disappear"? The patrolman ran back and forth – I am certain not knowing what to do "<u>anymore</u>" than any of us there. In about five minutes, he left the area – but never went to the field.   I had a front row seat to this unfolding right in front of me.

For this chapter, I would like to recommend a couple of books before I begin to discuss "portals". The first book I recommend to readers is called, "<u>Weird Kentucky</u>" by Jeffrey Scott Holland. This book explains many unusual things living among us here in Kentucky. The strange lights and orbs, ghosts, etc. It features many of our fine areas of activity – especially in Henderson County and Lake Barkley. The book in itself prompts other explorers and enthusiasts to come to our state to dig into our mysteries and report on them.

The second book is called, "<u>On The Path Of The Immortals</u>" written by Thomas Horn and Cris Putnam. I became interested in portals after a very specific incident in Carroll County, Kentucky. The exact location is not revealed to protect the people and the habituation site involved. Their book had the indepth analysis and reporting which will reveal great information/research/examples of things – we do not, won't, or cannot comprehend at this point in time in our history and in new scientific exploration.

The timeline <u>for this incident beginning</u> was in July, 2015. Also, I have had a few people (locally in Carroll) tell me that they believe portals are involved with "Bigfoot". Doorways, gates, and portals are things we the general public does not want to speak about.   We study them in earnest; but yet, we still do not understand teleportation, time portals, and other parallel worlds.   We do think about Heaven and Hell.

Does Bigfoot Use Portals? And, If Yes, What Do They Use Them For?

In July, 2015 – (Carroll County, KY) The night was warm and many observers had windows up to keep cool. This occurrence started at 2:30 am. to 3:15 a.m. in the morning. A group of them had gathered (Bigfoot) into a circle (basically) and whistled solid for 45 minutes. There was no breaks in this solid "whistling". They did not even come up for air. They maintained this same tone consistently (not a high pitch – just consistent). By it maintaining such a loud vocal, a local authority was called (the police) to investigate. As I think about "how that police officer must have felt" - pulling up and seeing this field a glow with yellow/orange color. It lit up the underneath of all the trees. The circle was about 20-25' across – with them gathered around it "whistling". I was not going out in the field. And, as I watched, the police officer did not go in the field either (which I think was smart). I think a bunch of people had assumed it was people partying and built a fire. I knew better and kept observing. I bet when the officer left in about five minutes he was thinking, "What the heck was that?" I have been thinking ever since – was it a portal, a ritual of some kind, or a <u>summoning ceremony</u> of some kind. Is this where their nickname, "The Whistler" comes from. I respect our police force for all their help (no matter what). That night, to be honest, I wasn't getting any closer either (since this was the first time I had observed such a thing).   That night in July, 2015 – I did not have a good enough night camera to do any photography.

So, I made a promise to get a better camera and wait for this to happen again in the habituation site. And, guess what….it did happen again on the last day of the month in September (<u>9/30/2016 – at approximately 11:00 p.m. that night</u>). I did manage to get a couple of night photos. The "glow lit up the underneath sides of the trees" and <u>there they were again</u>. However, no whistling was done by them this time. Observations that night: quiet, comfortable weather, no noise, just the glow this time. Also, no lightening in the sky was observed.

Insertion Of Glowing Circle: Photo Taken 9/30/2016 (11:00 p.m.)

Original Photo:

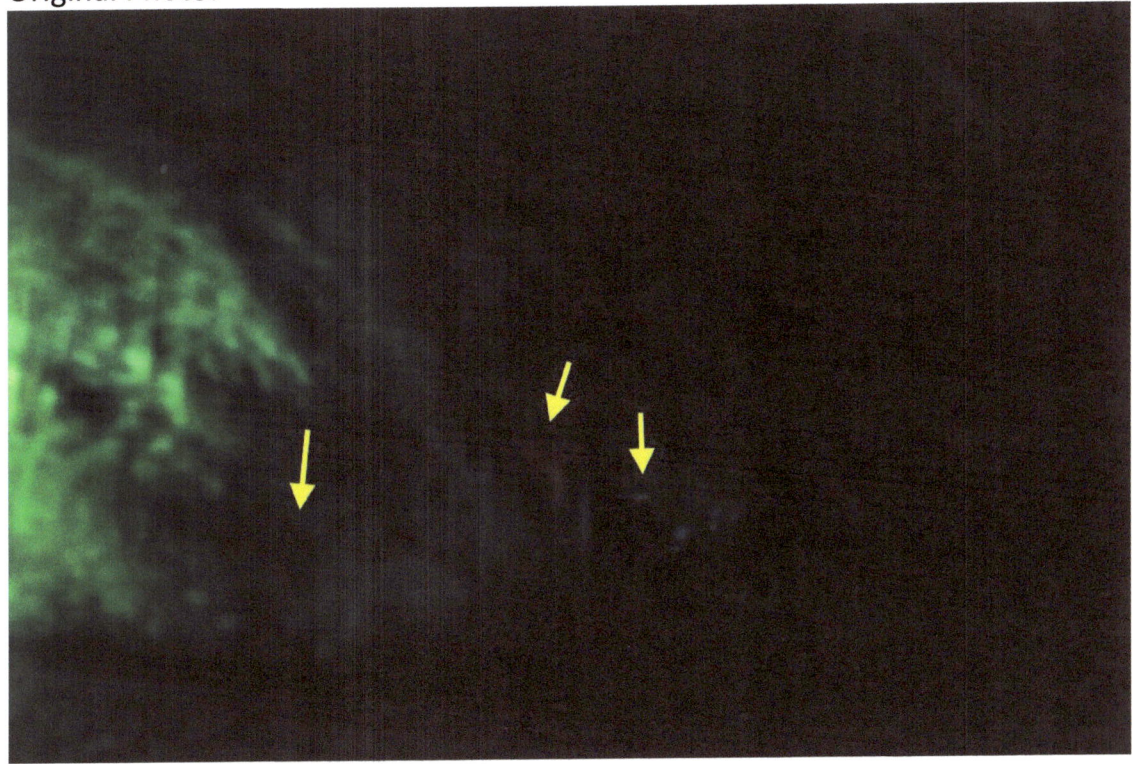

But, I did notice in the photo above a small "Bigfoot" sitting (on the right) beside this glowing and it is holding something. I cannot; however, tell what it has in it's hands. You will also see several sets of eyes to the right of the "Bigfoot". I counted seven sets of eyes roughly. To be honest, the small eyes or energies exhibited to the side is hard to ascertain for me as well.

It is indeed amazing to see this unusual occurrence and presents way more questions – than answers. But, we need to remember that every small piece of evidence could be a part of the puzzle (put with other researchers' evidence) to come to some kind of realistic conclusion (at some point in time later). I do know some scientists are studying ritualistic type behaviors of a monkey. So, this is good information indeed. What do you think as a reader? Was it a portal, ritual, or some kind of ceremony? The picture of the small "Bigfoot" was not visible to my eye but showed up in the developing process/review (which I was extremely grateful). I will continue to monitor the habituation site and see what develops further in that area. I do hope to see more and wish to receive more observations from this colony. Does it have a connection with a solstice? That was the last time I got to see this. It never happened that close to me again. It was a true "blessing" to be shown this by them and I am grateful.

(Breakdown Of Portal Photo)
Small Juvenile Sitting By Glow

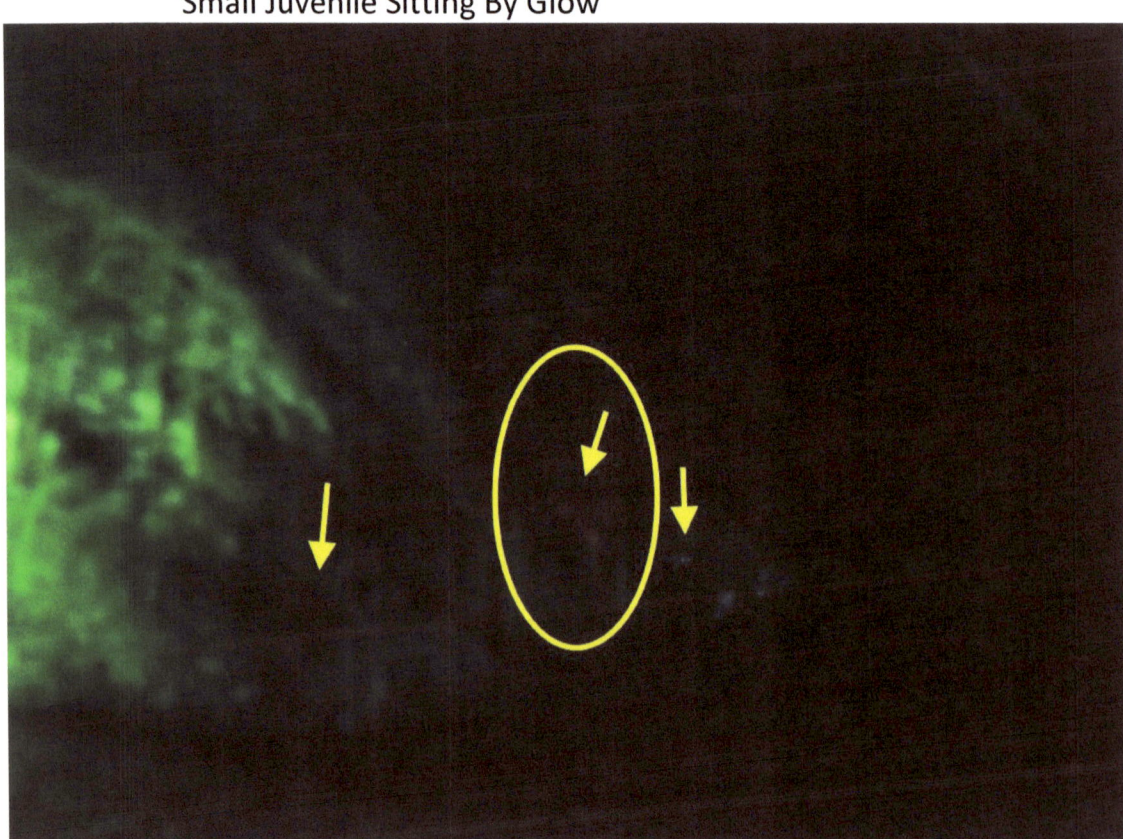

Continued Breakdown Of Picture
Large Bigfoot (see his hips, butt crack, vertebrae in back, head)

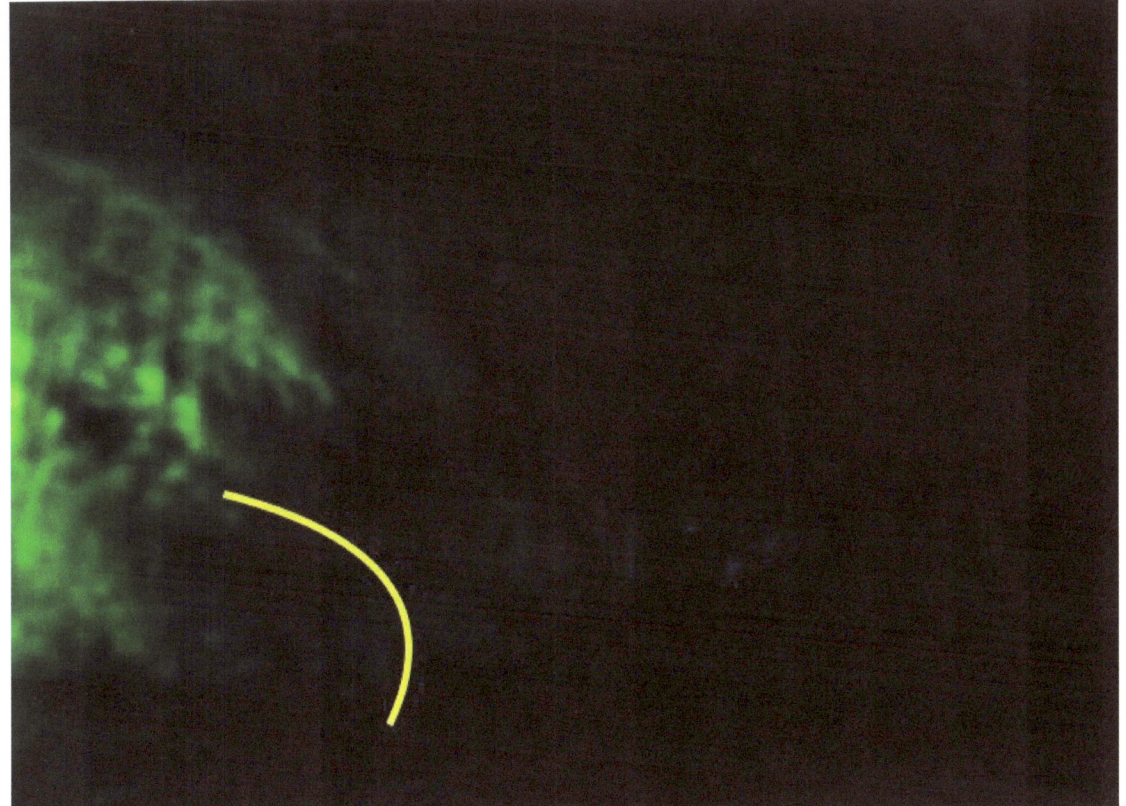

Chapter 13 – Babies

Additional Evidence:

This photo appeared in my book called, Kentucky Squatch'n, in August, 2015. I see three (small dots) on first observation; but I will enlarge this photo below so you can see what is peeking at my game trail camera under the log in the photo.

This is as it appeared on my game trail camera card. (I saw three small dots under a log.)

Photo Enlarged (7/16/2015 – 20:48 time)

The figure to the far right looks like a chimp exactly. He has black hair, beige colored chin, and his ears are sticking out on the side of his head. The second figure (looks like a small gorilla with black, thick fur, and his arms are a lot thicker). The third figure (peeking out) is unknown as to why his features are very unusual. He has dark colored hair on his chin and the top of his head (but white facial hair) and does not have a primate looking feature to him as the other two figures. These three do play together and it is clear they observe their surroundings in a team fashion. I do believe "babies of the woods" sleep together, hunt in circle, and also play together. I have a few very old, old photos showing them at this age circling and hunting together in a circle (to confuse the prey in the middle). I feel once they are able

to scurry fast alone enough; they are put on the ground to explore and grow both physically as well as socially in their hunting skills, exploring the land area they live in, and also learning how to "defend themselves". One example was seeing a small fawn enter into a creek area, it's ears pricked up as it was startled, and before you knew it – a hand was reaching out underneath the bush to "snatch the back leg". They had startled it so much it backed up to a bush. Coincidental, nope. They learned it from the elders (mothers/fathers) of their colony. Was it a shrill whistle from the circling crowd of the "babies" hunting to disorient their prey? I think the answer is a definite, "yes".
I think the mother stays very close to them for quite a while in their life. (Maybe five years like other primates?)

Photo taken In Carroll County, Kentucky on 7/18/2015 (18:54 time)
The "Baby Nest" with a game trail camera. (The baby sleeps peacefully).

*Notice the chin, lips, nose, and eyes closed.

The picture of this baby shows us a little more about their behaviors:

a.) They cover themselves with leaves and sticks to keep warm. I have seen a nest on the ground (built out of barbed wire taken down). This is then covered with leaves, a small opening left to get in, and lined with leaves (looks like a small igloo).

b.) They can be mischief makers (as I think this is the one that threw a small pebble at me to play or get my attention one day).

c.) I have seen this small "baby" air travel. I noticed when it rains or thunder strikes – it uses it upper body strength and swings from tree to tree. This was observed about the same time I got this photo. It was trying to get away from the storm coming in and heard a very, loud racket in the woods (eeking). I looked up and observed this happening. Sorry, was not able to get a photo he was long gone. It happens to the best of us – a missed moment on film.

d.) I do believe more than one sleeps in a nest at a time (the additional body warmth would help them). This is a great disguise for their safety as well. I am sure this also helps them to hide – if trying to learn to hunt and grab something (like a cat, squirrel, bird) when they are small. I have taken a photo of one reaching up out of a nest to stroke a cat's head. I believe their hands of comfort to our domestic pets are sometimes taken as "stroking" when, in fact, they are luring food in (training) to hunt. I am sure this is a learned process.

e.) They do build nests in trees as well (when leaves are on). I have seen them take two branches and twist them into a small nest (this is hardly recognizable by our eyes) – we think it is just a very full branch of leaves – with no meaning. Whether they sleep all night in them or just lay there to sun themselves, they sure got a good front row seat to anything coming and going around them. They do a long whistle of danger – if you go in their area and they want to alert a parent. I did this recently; it was very clear I was not wanted in their living room (of the woods) by a creek. I am sure I got checked out by a parent – but no worries, nothing aggressive happened that day. Just the all alert was given (a very long whistle).

On 11/12/2015 (at 2:07 p.m.) I took a picture of a Bigfoot'n friend researching with me in Carroll County, Kentucky. She told me after crawling into this hut that it looked like a little apartment building (where they had laid down). This researcher found about four to six sections inside. Also, we found a stick propped up against the opening to disguise the front opening to this hut. This hut is a very solid, weather breaker for them and and I am sure they gather together for warmth. She enjoyed seeing this for the first time; and we continue to "Bigfoot" together in Kentucky. She cast her first prints that day and helped me find other evidence.

This hut (if you were to see it out) – most people would walk by it as just wood trash – with no meaning. I think the stick is put in front of the hut to disguise the entrance; but also I think it tells them – if it is removed that someone or thing has been in their home. Some asked me the difference between stick formations, huts, and nests. I am showing examples of them in my book and also will be showing a t-pee structure. Some people have used this verbage interchangeably – for example, hut and t-pee. Just depends on how technical you wish to be. I always photo op the structure so you can see what I am calling it or not.

This hut or nest (at least 5 to 5 ½' tall) and (at least 8' long). It has plenty of room for the young colony (part of them to sleep in). Remember, when it is cold – sure they are in solid structures for warmth and collectively huddling. In warmer weather, sure some young one's will be lounging under trees, up high in their twisted green leaf nests, and older ones will build strong vining nests out of vines in the woods. My book cover on _Kentucky Squatch'n_ is proof of that. It was at least 50 to 55' above the ground and at least 10' long and 4' tall (vertically). So proud I found that one, it would hold a couple hundred pounds of primate in the air with no problem. He/she could see you coming a quarter mile away – head's up – way before you even looked up in the air. We have a tendency to walk with our heads bowed down (not looking up). We need to be cognizant as researchers (not everything of importance is at our feet) or on the ground. Our tendency to look straight ahead and down – is one hard thing I am trying to correct. I think they will climb as long as a tree is able to hold them up. I did have a picture at one time showing a total gray colored one. Very young and spry – climbing a gray colored tree. He likely was about 5 ½' feet tall – had very strong muscles – almost 200 lbs. His hair was a half inch all over his body. It proved to me two things: 1.) Not all gray colored Bigfoot are old; and 2.) If a tree can hold them, they will climb it. I could see his hair was the same length all over his body. I could see every detail of his almost bald head -his back was to me. His muscular back, vertebrae on his spine, and muscle built legs. His arms and legs were so strong – climbing in a hand-foot (right left style) – like what we see in cartoons "Spiderman". His foot was definitely adept at grasping the tree trunk. Believe it or not, I was so excited about this data and photo – I cannot find it now. If I ever do, I bet I have spent 200 hours trying to find it – is it gone forever? Not sure. I may find it again someday. I just wanted to show you – even I make mistakes as a researcher. It is all in the learning process.

They also will take fences (you will see many logs/trees rolled over fencing). I believe the older one's do this for younger ones. Yes, they can go over a fence without a knick – especially if it is a barbed wire fence lined on top. I also find many sticks just poked in the ground – about 4" in the ground. Many of them do this – why? Not sure – what this marker indicates. I have found four or five (maybe more) in one site. This is something we will have to study further. Also, I find they do not like new fences, especially if it cuts their normal "route of path" through a property. This is especially true if they want to stay hidden. Would they annoy your outside dogs and live stock and push on them to get back at you? Possibly. I feel (if you do something to annoy them – like cutting off a normal path they usually take) it is like a neighbor you had for twenty years. If you built a fence without telling your neighbor, wouldn't they be miffed. Maybe do a little mischief? I think so. Could they annoy your dogs to the point they want to break the back door down? Maybe, maybe not.

Photograph of a nest (had about six compartments inside for them to lay).

*I believe this researcher went out maybe two times during the summer months (she got her casts, got very busy, and did not return to the group). Some people want to try it and don't want to continue on. Others put the effort in a couple of times and they have their answer. Each person feels a definite way in their own mind how much time they want to devote to this "study/research". This lady is an avid hunter, outdoor's person, hiker, camper, kayaker, and explorer. She is tough for the outdoors. I look forward to doing more expeditions with her soon anytime she wants to pick back up. Her knowledge of the woods/tracking is of great help to me. I always want to learn from others as much as possible.

Remember, listen and learn from other researchers …………
My researcher was standing in a sink hole and we were measuring how tall this was (the vining, examining for prints (hand/foot), and other unusual finds that day). Remembering to …. look overhead…

Chapter 14 – Special Photos

This next chapter deals with social behaviors uncovered while researching. I find it one of the most interesting chapters for me. Why? It reveals to us a small glimpse into their everyday lives. Who do they hang with? What scares them? Where do they lay? How is the colony put together? Are the members of the colony photographed over and over. Are there newbies introduced into the colony? How much do they get around in the daylight versus the night? Contrary to belief, I was quite surprised from my photo ops to learn they travel quite a bit during daylight. The young ones often pair up together and lay down to sun themselves in the summer underneath a huge shade tree. I am sure all this "cohesiveness" like with other primates results in many cleaning, grooming, feeding, playing, and (yes) fighting amongst themselves learning to take over their new spot in the colony. This could be a young male stretching his authority or a mature male not ready to ease out of his power in this colony. He may think, "Does he have the strength, intelligence, and being a lady's man-would this constitute a dual?" I do think the juvenile that is wanting to mature and take the old man's place can be a threat to the family. I notice mother's more in the back of the scenes being protected. That does not say one might not "fight" to protect her young.

I did observe one evening a strong male (beating something on the ground). He hopped up after another "Bigfoot" signaled him; I was there observing. His face was gorilla like – very black, shiney (looked straight at me and left). I was unable to get a photo of this – too quick. He hopped up a hill and back into the woods. He was using both fists to take into hand what was on the ground. Not a deer. He was "blasting with both fists" the other "Bigfoot" on the ground; which now that I think about it (it most likely was a younger male testing him) to become a lead. I mean, he hit him with a "left-right", "left-right", "left-right". And, it was done. He stomped off. His facial expression was that of a "regular person - mad, angry, and in a hurry". I was in a safe observation point; but scary to me. His hands were as huge as baseball mitts-coming down at the creature on the ground. If he was killing a deer, he could have broke his neck and hauled him away. His m.o. was to take something into hand; he wanted to prove he was the big guy in this colony. If I had not interrupted this, I wonder how long he would have continued this beating session. Would it have been to the death or to injure him enough to "finally quit being the leader"? I will not know as I was found out spying on them-and the other "squatch" signaled him and he stomped out of there. Very interesting behavior indeed. Was jealousy a factor of the colony or just wanted to be number one? Was it for breeding purposes alone? How does he eat with the group? Does he find food as a leader and bring it back to them? Does he eat first and what is left-is left-and everyone tears off chunks of raw flesh and consumes it? Lots of questions still to be answered.

Each one of the items presented has great significance for me. I look for what the picture is trying to tell me about "them". I hope you do the same as I do when you view them. You may come up with a different meaning. This is quite alright. I am just stating what I felt at the time the photos were taken and on my review of photos of game trail cams.
I often look at the photo in front of the one and behind it – to get a better idea of who came in before, who was there in that moment, and what happened at the end of the photo shoot. Sometimes, it gives you a great deal of information being able to read their facial expressions, movements, and also – what is going on in the environment around them with other animals. A startled deer with it's ears pricked up, etc. – can tell you a lot – which direction is the deer looking……..I think you get the idea. At first glance of this photo taken in Carrollton – you will think there is nothing to see at all. However, as a researcher, this very nest has <u>yielded</u> quite a deal of information for me and I continue to study this very nest today for that reason. You would think that a small nest (where you had been finding a little – baby bigfoot) wouldn't bring in a much larger subject to "look at". But, I have found the bigger ones' do come in to check out the nest. They often will pick up a little baby, check on it, put it down and exit out. Also, while they are doing that in the nest, they will stand there and observe everything around them.   Possums,

raccoons, deer, cats have all strolled through the nest as well as a coyote or too. The protective parents are inquisitive and come to find out what's happening and possibly as a protector of this baby in the colony will visit many times during the night. I also think it is not just the original mother that checks on them as well. I believe it could be another member of the colony as well checking.   Their protector or parents also shows many examples of hunting from the nest at a very young age (even when they are laying down). They also push them over and out of the nest; I have a photo of one snuggling too hard and pushing a baby out of the nest (which makes the baby wake up and cry) with his mouth wide open. This photo will be shown later in the book.

Photo #1 (Taken 9/13/2015 – Time 05:06)

You can see this "Bigfoot" is showing his teeth and is looking straight at the game trail camera. He was approximately 15' from the camera. I believe he may have stepped in and adjusted it a bit – as the photo is crooked (like he manipulated the game trail cam). Often, they get ripped off by them or bears or human thieves (but, you just carry on and keep on bigfooting). As the initial photo is hard to see the face, I am going to try and enlarge in the second photo below it. At that point, you may be able to see his physical facial features better. He is most likely 7 ½ to 8' tall. Blue Arrow – Indicates (Top Of Head). Yellow Line – (His Teeth).   He is standing in the front part of the nest.

You cannot see the nest on the ground in the dark. It probably covers about a 5' x 5' space square (that the baby lays in). Not very big at all and this nest is not back in the woods. It is close to civilization in Carroll County, Kentucky.

We, as researchers think we need to cover hundreds and hundreds of acres in the woods to find something. This is not true. If they feel comfortable with you and there is a woods close by; they can go back to deep woods for safety. They are notorious for coming in. This nest was built out of wire fence, covered with leaves, about a foot tall and a foot and half wide. Just big enough for a small one year old primate to stay warm in. The wire was formed in an igloo fashion and also the leaves that lined the bottom – would keep him warm and safe, hidden (for a while). As I say, I believe even if a protective parent is gone for a short while (say to hunt) they come back regularly to check on them. From my rough count and what my photos show, I believe there could be ten to twelve in this immediate colony. I do know they come back pick them up, cradle them, and put them back down and go back out again. A final check like any good "parent" would do.

`Photo #1(Blue Arrow-Top Head) - (Yellow Arrow-Teeth)`

Same Photograph #2 (Enlarged and Marked):

You do have to ask yourself-what differences have you seen so far in the research from anywhere else?

Photo #2 – (Enlarged) What are the differences so far that has been seen in the colony?

<u>Male #1</u> – Brownish/Orange – has beige (primate muzzle) with black lines under his eyes. Partial hair on face, cone shaped skull. (See Page #11)

<u>Male #2</u> – All Black – no hair on face. More of a flat face and flat top of skull. (Pic below.)

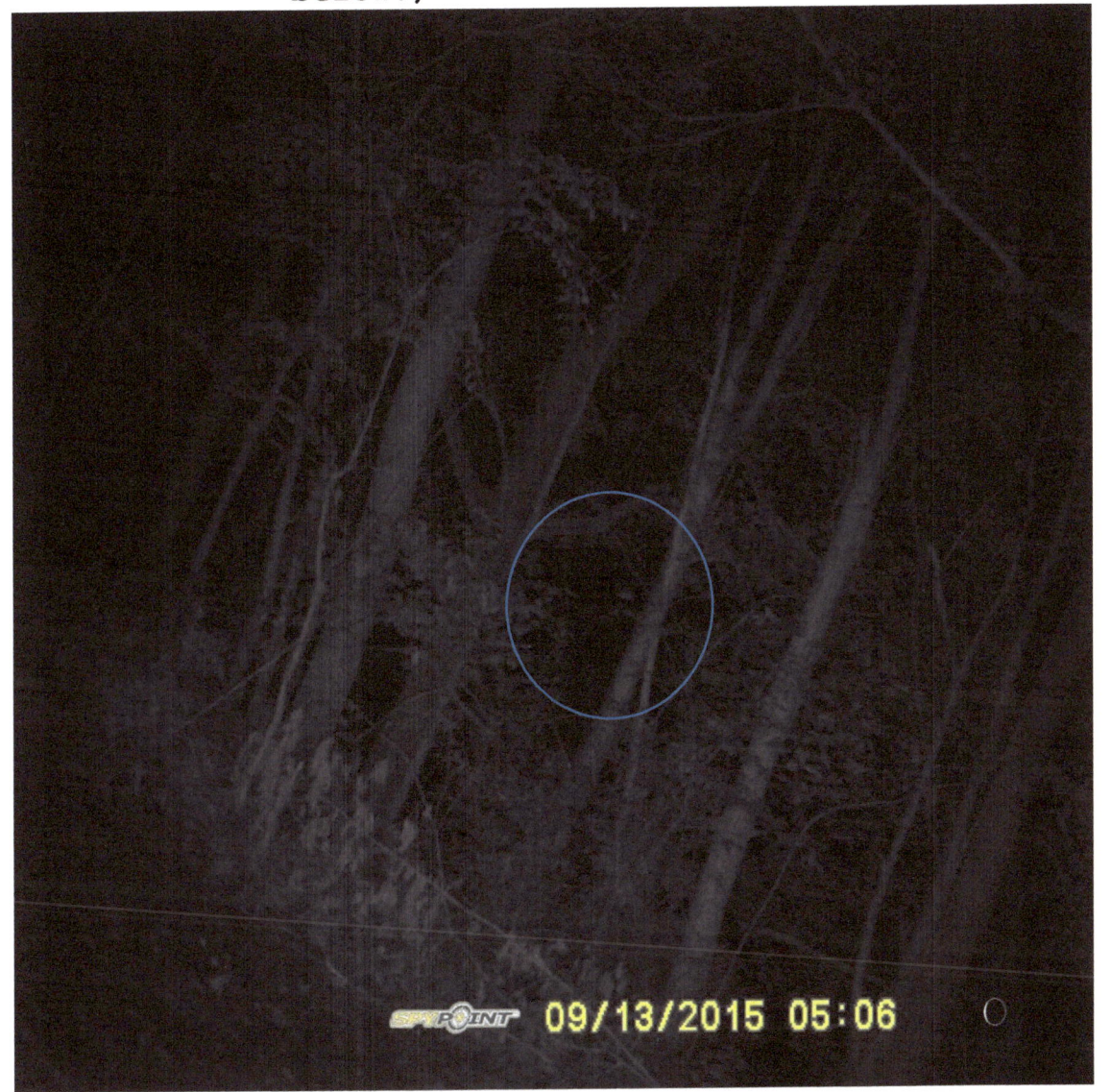

72

Photo #3 (Taken 7/19/2015 – Time 07:24)
Titled: The Crying Infant

(Left Arrow – Big Nosed Guy)   (Top Middle Arrow-Ear of Infant) (Bottom Arrow – Is A Leg Pushing Struggle – Like "your on my side of the nest".

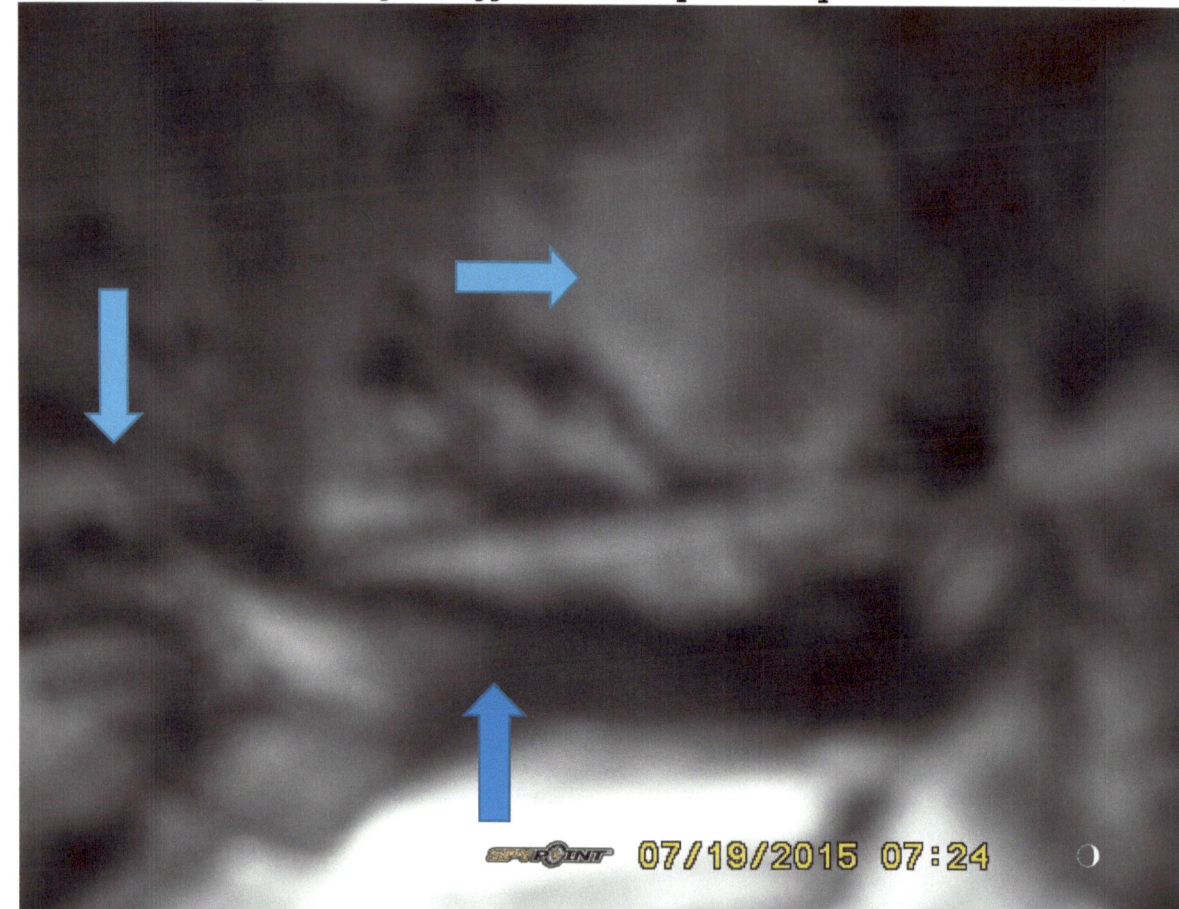

*Notice the mouth open. Small wrinkles on his face. This is a very, very small face – indicating to me (I feel) a newborn.

This photo taken in the month of July-2015 indicates there were several things going on in the nest in Carroll County, Kentucky. You see one arrow pointing to a face (which is crying-infant). He is either using his leg to push the "big nosed guy" older baby out and over or it is the other way around. One of them wants the other to move over and stop it. The infant is startled and his movements and facial expressions exhibits (one – he could have pushed over the other and woke him up), two – he did not have enough room to stretch out, or three a small animal could have came in and made both of them squeeze together. Critters roam in and out all times of the night in the darkness. His ear on the side of his head is very plain. He opened his mouth and screamed-gave a cry, you can see his chin line, eye sockets, and forehead. You will even see the little "wrinkles" on his chimp looking baby face. This one is very primate looking – beige skinned. I feel this is the youngest newborn photo.

Notice on the left side of the print, a big nostril/nose, chin, and lips. This is the second baby (lots older than the infant). The bottom arrow pointing (upwards on the big/nosed guy) indicates there is a struggle over this nest for room. As I say, this is my interpretation of this photo. There could be other interpretations from you as a reader as we go along. Nothing is set in concrete when researching. Sometimes, you think you have it all pinned down; then, shazam they surprise you with another new thing to start looking at. Like I said, babies of "Bigfoot" sleep, play, and learn to hunt together. There could be more babies in the nest and I just did not capture all of them in this one photo opportunity. Could they be saying (like our kids in the back seat of a car) – don't touch me? I definitely think so.

Photo #4: Baby Sleeping In Nest (7/18/2015 - 18:54)

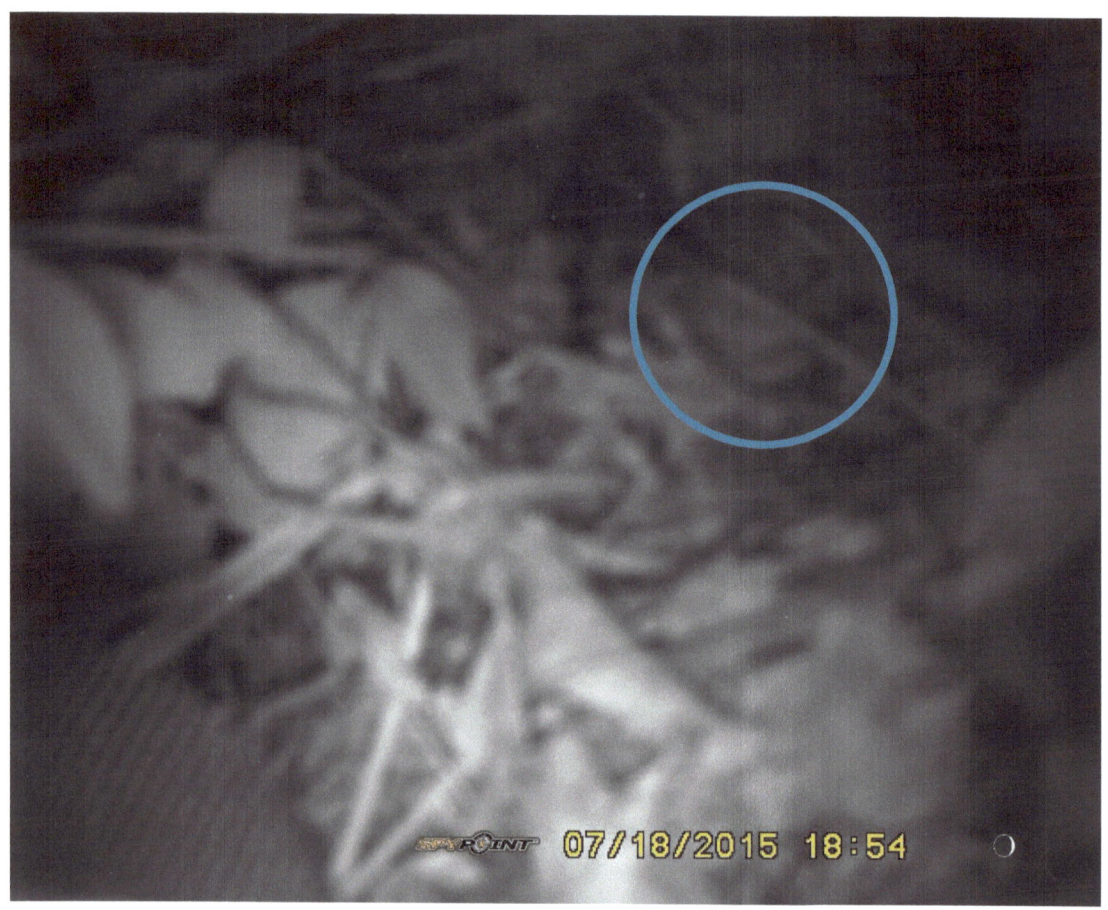

This photo shows that this little one was sleeping peacefully. I like to share lots of information on babies as people rarely get to see them. They are seen even less than the "big guys". They most oftentimes will play, swing up high in the trees together, and roam all over the place two by two (or even alone). Keep in mind, they look cute and cuddly but have the great strength at a small age to run extremely fast. Do not try to catch them or pick them up. They could rip your face off or an arm. I doubt anyone can catch them even our best olympic runners. They are afraid of rain and swing up high (eeking) hand-over-hand, tree-to-tree. I did see a young man try to catch one.

True story; <u>really funny</u> for me observing this scene. He was running barefoot and the "baby" had the advantage (even though baby was barefoot as well). The man chased the baby into a small thicket. The baby just kept on "moving" under the grass, leaves, and other floor coverings. <u>No Way</u> was he going to catch it. They are extremely strong, agile, can swing from tree-to-tree, run in a gallop (on all fours like a horse if they want to) when little. They learn lessons of life very early and that man is a <u>*threat*</u> to them even at this young age. He most likely was about a year and half old – very small chimp. Mischievous, wanting to play, and I believe he had scratched this man's window screen several times. The guy knew what it was. He was just not smart enough to realize–*what would he had done–if it turned on him*. This scene that I viewed was very close to the baby nest. I have a photo to present to you in this book, "showing a small hand coming up out of this nest". You do not see his head; but he is stroking the head of a domestic cat with his hand. Squatchers have interviewed individuals with regards to cats missing. I believe they are on their "diet" plate from a very young age. The lessons are learned from parents and other members of the colony. This training shows them water supplies, food, cleaning themselves and probably other members in the colony, and most of all – being "fearless" when need be to protect themselves. They do not fear man at that young age – they know they have the upper hand. A baby can out run us, out-smart us, and do "have the advantage of climbing" in getting a great escape from danger. They are trained on smaller animals; and then as they grow – they undertake hauling in bigger prey (deer). They hunt in circles and envelope the kill when young.

In some parts of the world, monkeys crack nuts with rocks or sticks. They use weapons (sticks, rocks, bones, etc.) anything they can to break things open. They will tear knobs off an out building, pick up small toys from peoples' yards, and barter with them. I have examples of this to show you later. I have had an empty turtle shell, a old roll of film, a deer skull found in a non-hunting area, a baby brush, and a homemade item (a vine strung over a log with two bark attachments - like a chime). They are smart and want to play. I encountered this very young one in my research and should have played with him. However, he threw a pebble at me. I was afraid at that point in my research he might pick up a bigger rock and toss it at me and hit a vital "spot" on my body. So, I did not pursue that. Now, knowing what I do; I most likely would try to play with him and gain his trust. That should have been my response as a researcher. To gain his trust and see him more, play with him every day, if necessary. "The babies and the parents can read your internal workings - if you are friend or foe". They can tell "why" you are there and they also notice if you move anything, take anything, or hunt in their area as well. Their hunting grounds and baby areas (where they raise young next to creeks/ravines) are home bases. They may run you out - the parents. Always make sure you can make a fast, clean, getaway if need be. Don't think you are going to figure an out when it happens. You do not have time at that point. If they exhibit aggressiveness, get out.

I do carry a .38 revolver now; but not with any intention of harming them at all. I do; however, do not run as fast as I used to (even at all). Due to coyotes, pigs, wild dogs - a person must be able to shoot a warning shot. We also have bears here now as well. Some friends have also ran into pot fields. It is a good thing to protect yourself for what comes.

This one baby (in particular) and I am sure now threw the pebble at me to play. I thought: 1.) would he let out an "eek" or "long whistle" to alarm his parents to come back pronto, 2.) would it be viewed as "play" or as an aggressive move, and 3.) I wondered if I did throw a rock - would he pick up a larger rock and throw it at me. I hear they can hit a target "dead on" with rocks. Also, some primates use rocks for medicinal purposes. I viewed a film of a very small monkey species picking up a rock (three times larger than his size - smacking it and sucking minerals out-dust) from that rock. He also did this because there was not much water in this one area - so breaking the rock and licking it took care of his physical needs. Mothers especially teach young one's how to crack nuts with a rock. So in one project, I put peanuts out and cracked with a piece of wood to imitate this behavior. To display that I understood and was not much different than them. I also was bartering with them.

I have had several things thrown at me or hurdled down a hill - to say "move along". The first was a branch smacked on the ground to get out of my area. This was in the beginning stages of research - camped out too close in the "gifting area". The second occurrence happened in Trimble County, Kentucky on an expedition. For forty-five minutes, I got stuck on a hill - and a rock slab - about 12" long, 1" thick, and about 8" wide was picked up - and flipped beside me. I know this _definitely_ was a message to move along - hurry up and get up that hill and out of my territory. I kept feeling like I was being watched the whole time I was at this waterfall. Please note: this was not the only time "I encountered a reaction from the woods – that I could not actually see who it was or the creature that was throwing it." This is Bigfoot's m.o. Little ones will build small "tee-pees" out of sticks and learn to leave their marks. I have viewed some of these structures built only about 12" high. Adults can push 30-35' logs (stand them up – into the fork of a tree). Some researchers feel this flagging indicates (a directional sign/plentiful food sign) and mark it.

Photo #5: Baby Face (8/1/2015 – Time 18:54)

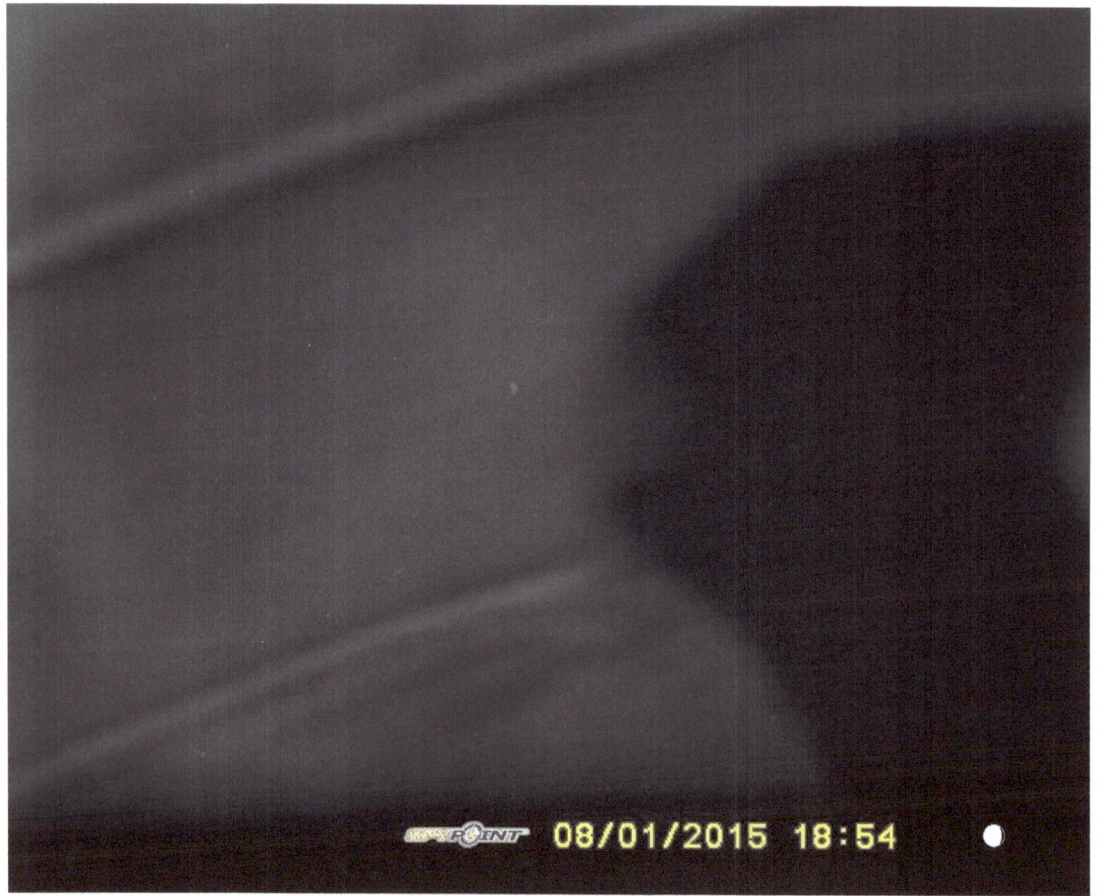

This photo was included just for kicks (this subject was close enough to the game trail camera to pick out the facial features/outline of it).

The definitions needed to say what it truly was – is not readily identifiable. However, it was taken within twelve feet of the baby nest. (Mouth Open-Cone Shaped Skull.) Sounds familiar, doesn't it.

Bigfooters report an electromagnetic field of squiggle type lines of interference with their cameras. They believe that is what caused this to turn out this way. The energy from the creature causes a camera to blacken to (nothing) here. I have also encountered photos that look like an old time t.v. screen where lines go across it (as static). What I find so interesting in this photo is the subject is yelling, mouth open wide, but looking at his crested shaped cone head – stood out to me. This is not your everyday ordinary creature from the woods.

Photo #6:   Unknown Crytid   (7/18/2015 – 18:53)
(Go to Page 80)

This photo shows something – but what it is I really cannot figure it out. I am noting it here so in case this thing comes back to the nest. I can then make some kind of prediction here at least. For now, enjoy the unusual photo – but this thing was sleeping where the babies normally huddle together in the nest. It has an elongated face, looks like antennae on his head, brown hair, and it looks like it is sleeping with it's tongue out. I hope to see more of this creature's features to make more of a "determination" of what it is. Hopefully, he will return so I can get a better look. (Final Note: he did not return to the nest.)

Photo #6 - Unknown Cryptid

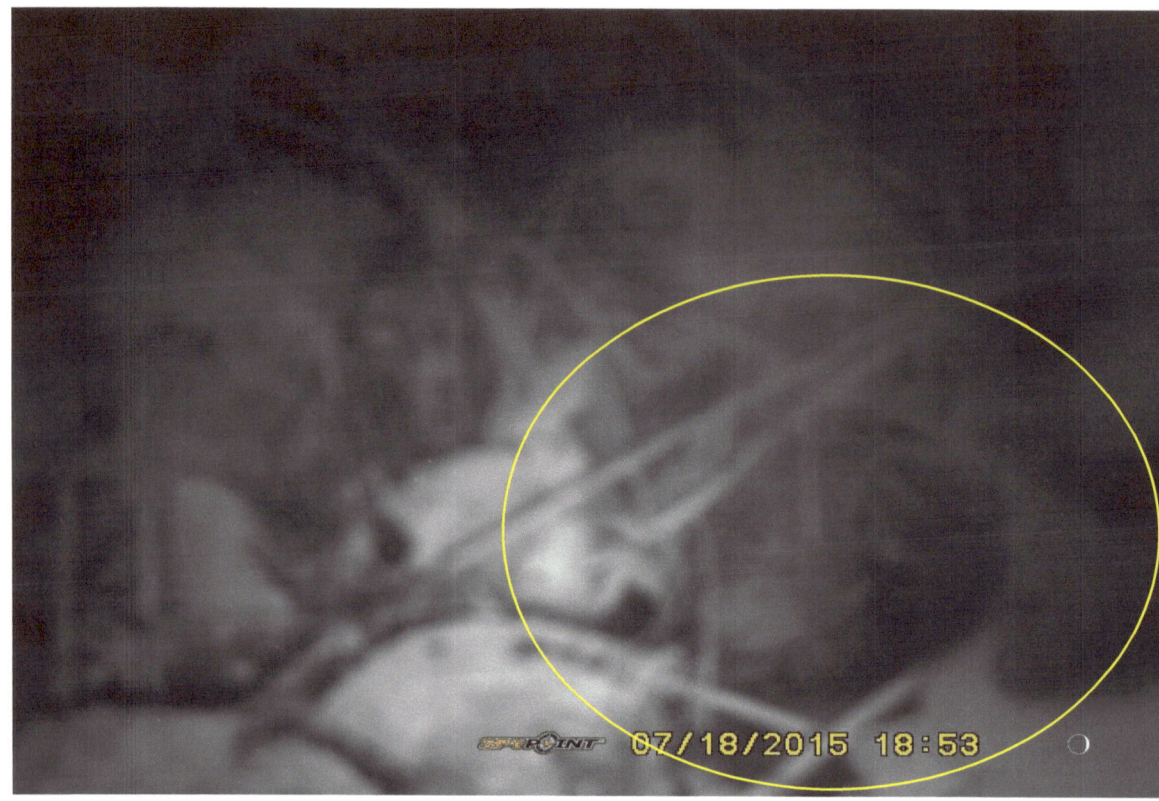

This creature is still not identified as of 10/2017.
He did not return and no further photos were obtained.

Photo #7 - Three Babies In The Nest   (7/16/2015-20:48)

These three were looking directly at my game trail camera. It was approximately 15' from the camera (is why the little chimp) looking one came through real clear (far right in photograph). I put this in my first book as a bonus picture as I was so proud that I had captured this "phenomenal photo" of a baby pretty clear and was proud of it. I have heard very few comments on it; however, but feel not enough people have seen it yet at this time. The mystery is, "Why do they all look different?" I would think they would all have similar looks (with their ears on the side of their heads, chimp face, black or brown hair and pretty routine/similar). However, this is not so. The middle subject looks like it has extreme hair all over it's body (like a gorilla looks). His covering is much thicker all over his body, black and very stocky - muscular build. Almost as though he is encumbered by the weight of his coat/hair. His arms is much rounder than the chimp looking one (with more defined arms/muscles in those arms). The third one, which really "whacks" me out, looks nothing like I would suspect playing with the other two at all that time of day. His characteristics are white-elongated face, dark moppy-type hair on his head, and dark hair on his chin. White hair covers his entire face.

Photo #7 (Three Babies In The Nest)

At first, I was not sure it was true about interbreeding of these creatures; but now, I am sure of it. Their physical appearances are very different. What could they be breeding with as each looks different? Are they breeding with other animals because we are squeezing them down into smaller and smaller circles of land/woods. Our woods are necessary for their survival and mating. Are true blood lines being corroded and dwindling – should the big males not be able to travel out and migrate and produce in other areas. Presently, it was interesting to find out there is a cross right now being investigated in the east of the United States of a wolf and coyote. This medium built animal is not as large as the wolf and not as small as the coyote. What do you think about this? Would love to hear opinions on this from my book. There are no right or wrong answers as we all research together to "resolve these mysteries". I listen very intently and as close as much as possible to others. It just may help me to understand and decipher something in my own logical – thinking type processes and theories.

Or, as many theorists are working on – as well as myself, "Is "Bigfoot" a creature that possibly was plucked out from our planet and experimented with by "ancient aliens?". Could cloning had been performed? Can they be cloning today as well? As a theorist, "I say yes". I know this rocks some people's world right off their foundations – many times over. But, we do have to consider this. Scientists are finding more and more – concrete evidence from "visits from Extraterrestrials". I took photos this summer of a UFO in the Carroll County, Kentucky area. Due to this finding, I am not sure all photos can be shown in just this one book. It is packed now with things. "How much more could I possibly squeeze in this book as research, evidence, and for anyone's examination."

I do stand by each and every photo depicted. All are legitimate photos. I never doctor any photos and they most certainly can be put under any archeological microscope, photographic testing, and I can swear to all locations, times, and facts regarding each one of them should someone wish to swear me in and do a "lie detector test". In fact, it would be most welcome.

I do believe cloning is taking place and has been done for hundreds and hundreds of years. If you ever hear of words like terraforma, which is seeding of a planet, or also - was the notion to continue their race - were their species that came to earth - dwindling. Was this experiment more for our survival or more for theirs? What was all the reasons behind this? More to come on my information and studies of this in the next book. It will be jam packed with information just as this book.

I feel too much thrown at people all at once - when they really don't believe in bigfoot to begin with...ya know - might be way too much for people (the general public to comprehend). We might just have even way more believers than anticipated. Many already believe.

Some of the public just want to see a picture of them and that is it. For me as a researcher, I want to know as much about them as possible, their purpose, why they are here, etc. We have way more to ask about, discover, and explore than ever before in our history of them.

Who knows - we may find more DNA in the future showing our genetics may be more (from the stars) than was made on this planet. That answer will come in the future years. A lot of people don't know that the DNA plucked out of our ancestry – is just a smidgen of what we understand about DNA testing. Many scientists regard the other information as "lots of extra stuff". And, there is a lot of "extra stuff" for our scientists and us to further examine. It could be a few years in the making or it could be another hundred years for us to decipher this complex mix. All one hundred percent of this complex DNA mechanism/model.

I am sure when the final answers come through – I will not be here on this physical planet. I am getting older and how much longer I will be researching – will depend upon my health and finding additional answers. I do hope my book helps new researchers so they do not begin all the way back at the beginning in research. This is lost time; I hope some very bright young and enthusiastic explorers yearns to discover things and picks up on the information left behind and go forth (young men or women and keep on finding out more answers). If each one of these people have to start from the beginning, there is a trial phase of learning. Some it takes, four years, ten years, and some in thirty years has not even captured a glimpse of the creature they call "Bigfoot". A most misunderstood race. I am very blessed that I have been able to view this creature, study it more, and write about my findings for others. I share these in earnest with everyone. I have had over ten sightings (babies and big ones) all at different times.

They can read your intentions. If you are friendly and go in with respect for them and the woods (which is their home), you will have a much better outcome than running and chasing them. Barter with this tribe just as if you were trying to gain trust with your own neighbor. Your new neighbor would not like to be chased down. I think if you build a fire, relax, and listen — they will come to you and most times will. Running through the woods is exciting; but it can also be dangerous for falling, holes, animals, and humans not trained to be in the outdoors. Be careful out there — as there is more in the woods than one might think. I have found that out as well.

This summer in the second colony that I am studying; I have already experienced some very unusual things. I took a photo and a gentleman dressed in the early 1800's appeared on one of my photographs. This same area yielded me a rock this summer with inscriptions on it. This, I believe, to be an artifact from the Shawnee Tribe which inhabited Kentucky throughout our state. I also took a friend with me that has the ability to communicate (telepathy) that I do not have. She picked up on several things: a.) the Shawnee was present, b.) this site was an area that a family had picnics often in, and c.) the "Bigfoot" advised her that when we come back — they will show us more. I found finding the rock very interesting as it is a concrete object. The rock depicted an area of "man or human life" drawn, a ceremonial area, and a "feather drawing — indicating Chief". All of this is so very, very interesting as it happened one, two, three this summer. This area is rich is lots of history, sacred, respectful happenings; and therefore, I did not remove the rock. I could not understand why I found it in about two minutes (turning over about three to four rocks only) and there it was. Normally, if you are a hunter of artifacts (fossils, arrowheads, etc.) you dig, and dig, and dig. I got it in a matter of a few minutes. Did someone want me to find it? A mystery to me today as I never gave rocks a second thought when Bigfooting. However, there is a message here — just cannot figure out what it is yet.

The gentleman I saw was riding a wonderful horse (strong and big) his front leg was coming up over the hill as he was surveying his property. I cannot believe I cannot pick this out of my films now. I have many, many SD cards and when I went back, could not pick him up again. His face, his hat from the early 1800's, his saddle, clothes all reminded me of a soldier from that era. Most wonderful, but lost in my archives of photos. Hopefully, I can find it again sometime. I remember how clean his face looked. Everything about him was clear and he most likely was headed towards the Ohio River area.

Chapter 15 – More Evidence

I had one gentleman on Facebook contact me wanting to know more about their nests, teepees, and wooden structures they leave in the woods. So Ray, this is for you if you are reading my book. I hope it helps you a lot in your trips to the woods (looking for evidence).

When I first go to a new place, I look for broken branches up high and most likely that it is the place to enter at that point. Also, when you enter, be quiet and listen for any knocking. Most times, they will give a knock to warn the "colony" and I have also heard knocks or a howl leaving. I kind of feel this is an "all clear" sign that we have left. They can see you coming from a great distance away. We have a tendency to walk and look down for evidence. Don't forget to look up, that is how I found the large nest about 10' long, 4' tall, and about 50-55' up in the air. Look in the air to see if the structure looks like something could lay down comfortably. The picture on the front cover of my book is a good illustration. This did not have any roads coming into or out of this area. This huge nest was a look-out point and they could warn anyone in the colony of me coming and going. The photo of this nest is on the cover of my first book, called "Kentucky Squatch'n". I was so proud to take the pic.

Also, teepee formations can be 12" high, built by small babies, or extremely large size (built by the adults) of the colony. The nests are usually built out of anything man leaves behind and the forest as well. I have seen: 1.) a wire fence pulled down and leaves covered over it, 2.) vines and wood put together as in a "igloo" style shapes – some are small to the ground, and 3.) the huts can also be shaped with a fallen tree root – example, if it leaves a hole, they will line it with leaves and put something in front of it. I have seen sticks placed in front of huts – as if a lock. They will line it with leaves and put in and around it for protection. This provides extra protection from the woods for them and also protects them from severe weather – cold. Babies do not like rain, thunder, or lightening.

As long as they can shape it, they will insulate it with forest growth which disguises it from our view. I often find sticks placed in front of the door openings. Is this a mark that if removed they will not come back to that hut. I always replace the stick back in front of the hut opening – so as not to indicate I was in there snooping. I am sure this tells them someone came in. If I find one like this, I inspect it and put the stick back and try not to be too intrusive into their home. I always treat the nest respectfully (as it is their home). Look up to see if any of the trees have great wear and tear on the tops of them. Their playing (the babies) will wear down and break off the tops of trees. Another clue as to some of their behaviors and how to track them in the woods, I have seen this lot in the area of Carroll County, Kentucky. You also see this in films of foreign countries – where chimps live – this happens here as well. They will twist limbs together that has leaves on them to conceal a small bed during the day. Oftentimes, you will see a big "X" made out of wood. This can be standing up or laid on the ground. Most times, other squatchers have indicated to me it is a warning sign to stay out. Of course, I want to go into that area – so it tells me there is a good chance of finding some information about them, evidence, and activity in the area to be investigated. Most times, I find a lot of walking sticks propped up against trees or fences. Man normally, if he has a walking stick he has cut down, takes it with him. In one area, I saw many, many of them propped up against trees, fences, stumps, etc. Could they leave them handy to do their knocking with? Makes me wonder on this one. I also found this summer (sticks about 1" around and 6-7' long jammed in the ground. If you looked at it, it looks just like a small tree taking hold. The difference is the jamming of them into the ground was always about 4" deep. I found about 5 or 6 of them in one very active "Bigfoot Area" of investigation. Why was this done? Not sure for an answer on that one, but I felt I needed to tell you about it. Very unusual find that day in the habitat site. What does it mean to them? One stick was left inside a stump. I knew that was there as it was used as a tool to dig grubs and insects out of this stump. I almost missed it as there was foot traffic of theirs around this stump.

They also will take (when leaves are on green) two branches and, as mentioned before, twist them together to make a bed up off the ground. Used mainly for the babies when resting. I really was fooled by this until I saw it one day. I would never have guessed they do that and I do not have a photo of this. Maybe I will get that physical proof on this one day. I just noticed it and left (my forgetfulness) on this one. Take a picture girl…it was by a fluke that I saw a small one do this. I am sure he could lay down there and blend in and I would not have detected him at all. Him being there and seeing what he was hiding in – brought it to my attention. I was so excited about the find and he got a good visual on me watching him – I did not get a photograph of it. This happened in a matter of seconds.

Larger juveniles will climb as well. As long as the tree holds their weight, they will climb high. I have seen a young one, about 5 ½' feet tall climb a tree that his coloring blended right in with the color of his hair. This is one of the mechanisms they use to camouflage themselves. He was not old. His coloring was "almost white-very light gray" and the tree color was the same. Anyone most likely would have missed this; but I caught it. They know what color tree to be in front of; and when little to stay low and up against tree trunks (where it is dark). If they are dark like the tree, they know this blends them in and I find several photos of almost all of them hugging a tree, their color. They are down low, hidden between two trees, under brush – makes it look like just a dark, empty spot. They know what they are are doing and the best way to conceal themselves from us humans/researchers.

One baby when I first started researching hid by going flat – chin out, arms and legs spread eagle down in between some rows of grapes (I was at a farm and winery). When I saw this, looked like a mud hole. But, to my amazement, he popped up and hopped over sideways. Not a dog, he had a black coat with a beige chin (chimp looking). I looked around and he was gone. Very smart little baby in the grape yard and his parents had taught him well. He knew where the goodies were to eat – it was right before harvesting the grapes to make into wine.

I will begin to show you some unusual formations I found while on my expeditions/tracking to different areas of research in Carroll County, Kentucky. All of them taken at different times.

Photo #1 - Heavy vining, thick, and not normal.

Photo #2 - Small igloo made out of very small vines, and lots of leaves. Notice the igloo shape in the front of the photo.

Photo #3 - One of my reseachers standing in a low lying area. We found footprints and cast them and also found nesting in this area on the ground. One area had a nest in the tree and it was rounded - liked a hammock (to lay in - pretty much as we do).

Photo #4 - Made out of fresh leaves and vines. Very weather proofed for them.

Photo #5 - Two huts in a row. Crawled in to find hair samples. We did not find any that day here. I crawled in and looked and looked for hair and other items. This was on the same property we found so many footprints to cast that day.

Photo #6 – They totally filled up this old cattle stall with vines and coverings. Great disguise-it just looks like a bunch of over growth in this. However, it is the same property. Great Disguise – just looks like a mess and nothing more. Pond close to this with a lot of prints I found around the pond. Know they go hunting, drinking from that pond. The nest is right up close to this water and is very handy for them, indeed. I have seen deer (live and dead) ones. Plenty to eat.

*Counted about 20 deer (small herd) in this area.

Photo #7 -   (X) Marks a warning to those that enter.

Photo #8 - Special Markings on a tree. I am about 5'2" in this photo. Please don't look how grubby I am on this trip/expedition - no bath, hiked all day, and pretty dirty. Had the most wonderful time on this expedition in Trimble County, Kentucky. Markings very high on the tree. This is where we heard grunts for about two hours in the middle of the night. <u>They wanted us out of their area.</u>

Chapter 16 – The "Dogmen" Encounter
(Wrote About In 4/2017)

On January 12, 2017 – (time 10:45 a.m.), there was a lot to do that day and I went to my honey hole expecting to see evidence of "Bigfoot". However, I just had a funny, uneasy feeling when I go to this place alone. It was about to rain; the temperature was warm about 65 degrees and cloudy overhead – as far as weather goes. A little warmer than usual for this time of year.

That morning, I put on my backpack/gun, and made sure I had my cameras in case something caught my eye. I left the car about 10:30 a.m. that morning and was done at approximately 11:30 a.m. and noted the time. I was out there less than an hour.

As I hiked to the spot (where I left them gifts), I looked for signs of tracks, trees pushed over, new changes in the surroundings. I really did not hear any knocks or anything when I entered the area.

As I was taking down the game trail camera, I noticed they had put limbs around my trail camera. Small forked branches leaned up against it as either to cover it or disguise it from anyone else. Also, I looked across this field leading to the actual forest and noticed a small group of animals (deer – five) running up the hill. I also noticed a walking stick leaned against a log. This was a good sign – they had been there.

The tracks I noticed were not very discernable. I decided to do no casting that day. In a very damp condition – the drying time would take forever. When I picked up the camera, it was about one foot off the ground. I had hid it low so as not to be seen by someone wanting my game trail camera.

I unlocked the game trail camera and gave it a wave with my hand - to see if the red light indicated a full battery charge. As I waved it, this is important; I had it tilted toward the sky (upwards) directed at the top of the trees in that area.

While I was in the spot looking around, noticed the wind was picking up and it was getting ready to do a pour down of water. But, before I left, I heard a knocking in a small building by me. Then, as I was walking out, two plain knocks - was heard. This signaled that I was leaving. I am going to start leaving my recorder run (voice on) everytime I go out from here on out. By myself that day, I made a goof. I should have had it turned on (but didn't). One of the other most important points as well, I heard scratching "like with fingernails on a board". This really creeped me out at that point - it felt dark, very dark - uneasy feeling in this spot. So I was getting ready to leave - just uncomfortable and not safe. As you notice by the times given, I was not there long at all. There was a canvas curtain over a window (green). Something was flipping up the tail-end of this curtail - but not visual to my eye. This also made me "turn tail" and want to get out of there.

Now, for evidence, in this section of "Dogman". I will show you several photos and how I feel about them. You will have to draw your own personal conclusions after the evidence is given. I cannot change any facts to my story and welcome any expert to send my photos for testing at any authentic lab, photo specialist, or expert they want to. I certify the photos were taken with my game trail camera on that date.

Photo #1: There is a large field approximately 75' to 100' you have to cross to get into this area. There is also an extremely large water source nearby. I am circling a small figure sitting by the lowest part of a tree across the field. This could be a "Bigfoot" juvenile due to his size. I do not think he is a fully grown mature one. The photo of this one by himself (is not as good as the photo of the colony in December-2016).

(Go to Page 103 - next - shows original photo).

Photo #1 - Original Size

Photo #1 - Enlarged Subject

Photo #2:   Notice, when I picked up the game trail camera, being held in my hand – it took a picture (of the top of the trees on the hill). As you can see, they are very tall trees, and they are not what I would call at the top of the trees <u>substantial enough</u> to hold the weight of a full grown, mature creature of any kind. This is an important fact to keep in mind. The trees were not bending. Also, I then decided after looking at the photos to see if anything else was in there to report and note. Please follow my results in the next few photos.

Note: Photo #2 indicates a regular setting on my game trail camera. There was no enlargement of this photo on the next page. I shortly started hearing some scratching and left about 10:50 a.m. that area that morning. There was a bad like, dark feel to this spot I had not felt before. For some reason, I felt I needed to leave this spot. As the scratching was loud, there was no visible sign of anything close to me in or around the area I was standing on. What was scratching? And, why could I not see what it was. No big scrubs or any kind of brush or grass. Just a big old lake that was close by and a cabin.

Before this event, I only looked for Bigfoot and nothing else. I have been followed and heard an extra step behind me at this spot; but I never had this dark presence around.

Photo #2 - Regular Size Photo From Game Cam 106

Photo #3: **The "Dogmen" Photo** (Enlarged Previous Pic)

I truly love this photograph. It captures a male on the right and a female on the left. If you notice, she is holding a baby very close to her chin. You see a very small head. This photo taken by accident (picking up my camera caught this - as I waved my hand in front of it to check the red light). I was checking the battery level before I opened it. I had tilted it toward the top of the trees - and snap; it took the <u>photo of a lifetime</u> for me.

I did not know what was on here till I started checking the photos. This was a quick snap of the game trail cam by a fluke when I picked it up.

The picture was enlarged when I got home. I saw two very big, black dots on the photo. So, I enlarged it more, and that is when it was clear I got a true "Dogman" couple on film. If I had not been meticulous in my research of this photo, I may have took a very short glance and moved on to the next photo from the game trail cam. I do believe some people are throwing all kinds of pics away as they rush through their researching of SD cards and photos. You really need to take your time.

The male (which I think is what they call <u>Dogman</u>) is looking out over the woods - as if surveying what is going on. Please take note, I did not see them. I was there 10:30 to 11:30 a.m. that morning and no signs of them physically to my naked eye. The male looks dog like and has a black coat of hair, shaggy, and a long, pointed muzzle (canine). The male has a tail.

The female, due to her stance, I cannot tell if she has a tail or not. I believe it could just be wrapped around her leg in back of her. The baby is very tiny held to her chest/chin area and is resting comfortably under her chin (her left side). I believe these two to be mates with a baby.

He looks very protective and her stance is very motherly with the newborn. How old this one is – I do not know. Will have to study "Dogmen" history and facts more. I do know there is talk that there are six (6) different kinds of them found. I do not believe the female is a squatch or bigfoot due to the fact I studied the pic some more. I do see part of her tail wrapped around her legs. I really have not seen a squatch with a tail.

I enlarged Photo #3 so you can see their whole body of each creature they call "Dogman". <u>Go to Page #109</u>. Was that what was giving me a funny feeling – hearing that scratching on a board sound – like a dog pawing a board with their nails?

(Dogman) Couple

Photo #4: All I did to this picture was enlarge it, you can see the female's face up close. You can see lines of her eye sockets, nose, mouth, the small baby (cradled under her neck-her left side), and both have ears on top their heads. How aggressive they are, "I definitely say yes." They did not chase me; however, I have heard people being ran after. I do have a theory on them now though. I believe they are "very ancient" and in line with "Bigfoot" as to some of their unknown physical features to the regular public.

They were not physically there when I was there. I picked up their image on the game trail camera by the energy from their bodies. They, like bigfoot, have the ability to <u>not appear in a material form</u>. Their energy must have been just enough for my camera to capture them up in the trees.

Here is my thinking and evidence. Also, note their feet and hands are not touching the trees or branches (just as if they are levitating in the air). The male specimen especially has no grip on any branch or tree. Their sizes (as large as they are) in the small tops of the trees would not hold them (they should have been pulling the tree tops over with their weight).

Also, when I was in this place a time or two before, I had an unusual feeling that something was stepping behind me (around a circled tree I was investigating). When I stopped, I could hear one step – stop behind. I think they might have been the ones following me that one day (checking me out). Every time I looked back, nothing was there. It kept up continuously; so I left that day and did not do anymore in that area (for fear – I did not know what would happen). Would they become "solid" creatures and chase me and I know that I cannot out run them?

Photo (Dogmen)

I would say these "creatures" were up in the air off the ground at least 60'. I did not know they were physically there. No measurements could be taken due to them being so high up. Another important fact, if you look real close at the female, you can see a "branch" from the background going through her head. She is not totally there in what I would call a physical presence. I can see the branches in the background as if looking through her. If she was solid, I would not be able to do this at all.

We do know from reports - "Dogmen" can be very dangerous and aggressive. I did not feel any threat (just uneasy) out there by myself that day. I am going back there some time in the future and take some of my "Bigfoot'n" friends with me and do a more elaborate search of prints, evidence, and anything else we can find on them.

For now, you will have to go with your conclusions as to these photos. There has been no tampering with the insertion of photos or figures. They are only enlarged so you may view them better.

All are actual footage, and precise in the facts and truth. I welcome any expert in the field of "Dogmen" to contact me and am willing to send my footage for investigation and the results they find would be posted in my next book as to their findings as well. They may even be able to pick up on something that I did not that day.

Could the unusual crytic beings seen in the forest all have a connection to their same origins? I do think the answer is a definite, "yes". They are protectors of their species and also of our planet, woods, and streams. I think one of the messages being sent us is about our environment. We must look to the "stars" for their connection with us on this planet and us. This summer, I have been also taking photos of UFO activity in the Carrollton area of Kentucky. I have viewed some of the most unusual things and in doing so, I will be publishing my findings, connections, and proof of these things in about two years from now. It will be a continuation of "Bigfoot Madness, Vol. II". I have been shown some phenomenal things and am still gathering some other incredible finds from the woods. Whether they are somehow pointing me to hone in on them or I have just happened upon these things, I may never know. I am truly blessed to have seen these things in the open field. As I am a true believer in "Bigfoot" and also "UFO's" and the species coming to our planet, I sometimes feel as though I was jettisoned or thrusted into the future. Living in the present day, very few see the "true connection" between us and the star people. We definitely have DNA markers - from them, I am sure, as well as our primate world has DNA markers from visitors to our planet. Scientists are finding more evidence to gravitate us to travel space and live there - from which we probably came. Lots of evidence is pointing towards Mars.

Why do they not make themselves (Bigfoot) present to everyone? Why are they aggressive to some and not to others? So far, I have been lucky to not be chased by anything in the woods and always try to be as safe, responsible, and respectful to the creatures I study and the environment. I do know these things can read your intentions towards them and do believe this is done telepathic style. If they trust you, they will play, leave you gifts, and show you all kinds of neat stuff interacting with them. I hope to concentrate on some form of language with them in the future. I have been experimenting with some things in hopes it pays off.

This summer, I held my first Bigfoot Conference in Sligo, Kentucky. At that time, I met a psychic medium which I took on a trip to the second Colony Area - I am studying now. We had a visual and noise of something black going over the hill the first trip out together. We found lots of evidence that day and I also came across a Shawnee artifact (rock) in a creek bed. She advised me that a young juvenile would sit close by the edge of the woods and watch cars go by. She advised me that (right before we left). The big one said on our next trip out they would show us more. The Shawnee Rock Carving (was left) but I photographed it - and after looking on some charts - I found out it signified mankind, water, boundaries, chief, and a ceremonial area in this spot. I left the artifact as I did not want to disturb it. What kind of floored me - was the fact I found it in about two minutes. I had only turned over about the fourth rock - and there it was, easy to find. I found that peculiar - normally a hunt all day might not turn up anything. Why was it shown to me so easily? There definitely was about a thousand rocks in this ravine and creek area of various sizes. It reminds me, a friend once told me after speaking with me about Bigfoot that I had "Bigfoot Karma". I am not sure what it is - but I like seeing them and studying them for sure. They always seem to leave me with more questions.

If I do run into them (when they are in solid physical form), I do know that they are aware that I am not aggressive and just studying them. I (even carrying a weapon for coyotes and other things - bears) know that they are aware that I would not turn it on them. I have never killed or wounded any creature.

Many claw marks have been found on vehicles in Kentucky and many of them cannot all be "Bigfoot" and "Bear". Could they be "Dogman" marks? I think some of the claws could very well be from them. I found this a most interesting trip out that day even though I was by myself.

It is better for at least two people to go hiking for many reasons. Safety just in general but you always want to figure if you (got a way out) just in case something thinks about "chasing you". Many an experienced hunter has ran into pigs, packs of dogs, and coyotes in Kentucky. You also want to watch for our hidden snakes (as they blend in way too well in our area of the woods).

The most important thing – I think if anything totally catastrophic would happen – most likely it would have happened way before now. To my knowledge, unless people are keeping mum about it, there are no regular killings or special human attacks. However, I always do go armed. I do know some hunters have been ran out and left all their brand new gear behind. How do you fight this thing should it become aggressive? I do not know the answer to that one. Also, how do you protect yourself from something that can be in a solid state and then just vanish?

I will take it a case history at a time each time I go out to the woods to research. My question to you, "Is the public ready for this?" I think they can handle it – the knowing of an "entity" not totally biological all the time. We do have to keep in mind we are not the most intelligent and it will require studies from almost every imaginable category of science you can think of to further do some of this specific testing. I do believe this specialized testing will come in the future (but all cases must be reported truthfully) to not lose the historical, factual reports of incidents. Some of this material may not be suited for very young children to understand at this point. Therefore, decisions to share with your children some of the reported findings is solely up to the individual parents. I have enjoyed sharing the research and do hope you will write to me and also share information with regards to your outside trips to the woods. Also, I found out the many etchings on cave walls from in the past depict a man with a "Dog's" head on top. Our ancient ancestors knew this creature very well – he drew it, named it, and left lots of factual information for us. You need to watch History Channel's, "Ancient Aliens" and become very familiar with the pyramids, Greek mythology, our star system with planets, and the many, many clues with regards to our ancient history. It all starts fitting together and scientists are already back peddling on some of the former firm statements they made with regards to man's origins.

We have been visited by ancient extraterrestrial aliens and they have inserted part of their markers (DNA) into our creatures. Why wouldn't they also insert/share some of those markers into humans? We still have many mysteries to solve given to us from the forest people and also from our visitors to our planet.

My next book most likely will come out in about two years under Bigfoot Madness, Volume II. I want to take my time to show you the UFO photos, the embryo looking subjects descending from the sky to earth, and also the propulsion photos taken this summer from April to October, 2017. All of the above materials is true and factual (copyrighted 2017)-special historical documents from Carrollton, Kentucky.

I sincerely hope you have enjoyed the photos and all the evidence that I have presented. I am going to close and wish each of you a great experience in the woods, a wish to protect our planet and environment, and teach our children how to carry on in the same fashion. It is extremely important that we protect not only our planet, water, air, land - but also to protect all of our animals on the planet. Our creatures, birds, insects, and amphibians are most important to our own survival. I will close now and leave you with a couple of bonus photos from the next book being worked on.

Bonus Photo #1 (Shawnee Artifact Found Summer of 2017) The meanings will be revealed in my next book.

Photo #2 - Propulsion From The UFO (Taken 2017)

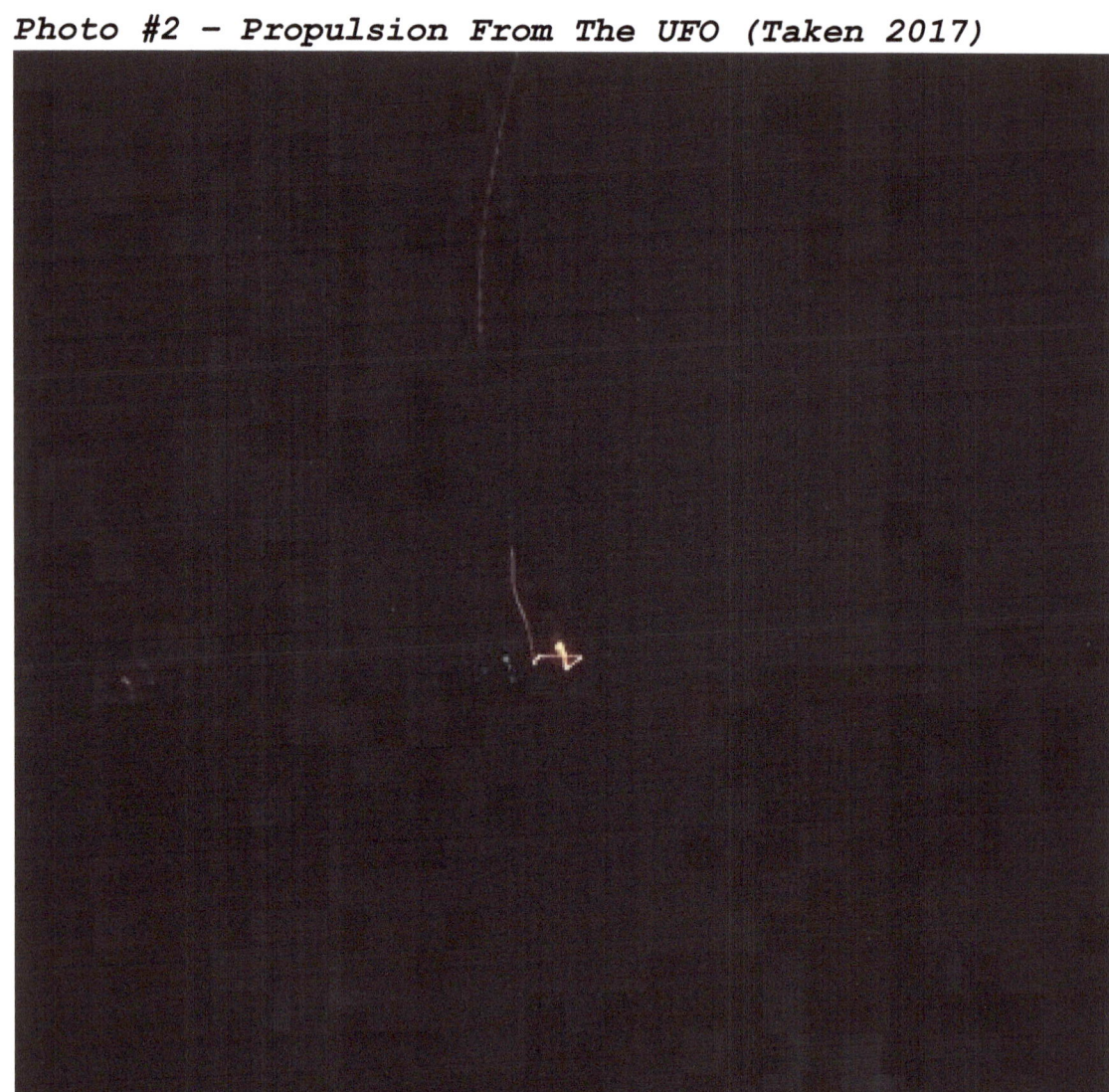

Photo #3 - UFO Picture (Taken Summer of 2017)

Reva McManis has been researching since March, 2013 and has lived in her local area of Carrollton, Kentucky for the past 25 years. She loves researching, writing, and bigfooting in her local area of Carroll County. She will get calls from other local areas with regards to UFO's, Dogmen, Skinwalkers, and Bigfoot.

Her experiences are insightful and you truly will enjoy this book showing over twelve (12) of the inhabitants of the "The Bigfoot Colony". This past summer in 2017, she experienced several new pieces of information that she is presently working on (Volume II) Bigfoot Madness to be released in 2019. The UFO crafts have gave her a glimpse into our skies.

www.ingramcontent.com/pod-product-compliance
Lightning Source LLC
Chambersburg PA
CBHW051151220526
45473CB00003B/734